T0282540

MAPS AND SURVEY

MAPS AND SURVEY

BY

ARTHUR R. HINKS
C.B.E., M.A., F.R.S.

FIFTH EDITION

CAMBRIDGE
AT THE UNIVERSITY PRESS
1944

CAMBRIDGE
UNIVERSITY PRESS

University Printing House, Cambridge CB2 8BS, United Kingdom

Published in the United States of America by Cambridge University Press, New York

Cambridge University Press is part of the University of Cambridge.

It furthers the University's mission by disseminating knowledge in the pursuit of education, learning and research at the highest international levels of excellence.

www.cambridge.org
Information on this title: www.cambridge.org/9781107699601

© Cambridge University Press 1944

This publication is in copyright. Subject to statutory exception and to the provisions of relevant collective licensing agreements, no reproduction of any part may take place without the written permission of Cambridge University Press.

First edition 1913
Second edition 1923
Third edition 1933
Fourth edition 1942
Fifth edition 1944
First published 1944
First paperback edition 2014

A catalogue record for this publication is available from the British Library

ISBN 978-1-107-69960-1 Paperback

Cambridge University Press has no responsibility for the persistence or accuracy of URLs for external or third-party internet websites referred to in this publication, and does not guarantee that any content on such websites is, or will remain, accurate or appropriate.

CONTENTS

PLATES

PREFACE

In the Preface to the first edition of this book it was explained that the work was designed as an introduction to the study of Maps and the processes of Survey by which they are made.

In planning the work it was necessary to decide in the first place whether to call it Maps and Survey, or Survey and Maps: that is to say, whether to take the logical order, of Survey before Maps, or the order of general interest and use, which is Maps before Survey.

Since many more people use maps than are engaged in making them, or than need to know the details of how they are made, it seemed better to begin with the consideration of the map as it is published and used. A clear understanding of the best results that have been produced up to the present time, leading to an appreciation of the ways along which progress is desirable, will make a convenient basis for the study of the methods of Survey, and the manner in which our representation of the world's surface may be extended and improved.

Some years of teaching Topographical Survey and the elements of Geodesy to the students of the Department of Geography in the University of Cambridge had shown me that there was need of a book to give a general account of the many-sided art of Survey. The official *Textbook of Topographical Surveying* by Colonel Sir Charles Close, F.R.S., is invaluable for instruction in all the details of the various processes. But it seemed to me that the student needs some explanatory introduction, unobscured by much detail, which shall exhibit the general nature of the operations, and the relations to one another of the various parts of the subject.

It is certain, also, that the student of geography requires an elementary account of the operations of Geodesy proper, that is to say, of the higher survey whose aim is to contribute to the knowledge of the size and shape of the Earth; and which has in recent years extended its enquiries into the constitution of the Earth's interior. These subjects are entirely excluded from the official book mentioned above, and I do not think that there is any book which gives a general account of this most interesting part of the subject.

The ultimate refinements of Geodesy cannot affect the maps in the slightest degree, cannot be said to be of any practical use whatever, can offer no return in cash for the money which may be, and which ought to be, spent upon them. Their justification is on a higher plane. A nation is judged, and rightly judged, by the public spirit and the public taste which it shows in its buildings, its pictures, and its gardens, which any educated man is, or thinks he is, competent to appreciate. And equally a nation is judged, though by a smaller circle of judges, for the contributions which it can make to pure knowledge. An exquisite piece of Geodesy may give as real a pleasure, and be as genuine a source of pride, as the masterpieces of art and literature.

I made no attempt to describe the minutiae of instrumental adjustments or processes, believing that a general view of the subject should not be obstructed by a mass of detail which is tedious to read, but had better be avoided until the student comes to deal with the instruments themselves, and carry out an actual piece of survey with them. Nor was I able to give anything more than the slightest sketch of the large subject of cadastral survey. This is an intricate subject whose methods are to a great extent governed by the system of land registration and taxation in force in the country to be surveyed.

While the literature of Geodesy and Topographical Survey is extensive, comparatively little has been written in English on the subject of topographical representation on the map. The subject is relatively new, because until the introduction of colour printing there was not a great deal of scope for enterprise. It is still in the experimental state, as is evident from the continual change in the style of the publications issued by the principal map reproduction offices. In these circumstances discussion and criticism are doubly interesting, and I devoted more space than might seem necesssary at first sight, to the detailed analysis of typical sheets produced by the leading surveys of the World. But since it is impracticable to give adequate specimens of these maps, this analysis can be effective only if the student is able to study a selection of the actual sheets. For this reason I gave the sheet numbers of good specimens, in the hope that the schools of geography who may be interested in the subject might find little difficulty in procuring a characteristic series of maps.

In the years of war that followed the publication of the first edition many changes were made in maps, and many interesting processes of survey were devised. In a second edition, published in 1923, these developments were treated in new chapters, the bulk of the book being reprinted without change. The second edition was therefore transitional from the pre-war subject which I had taught in the Geography School at Cambridge to the considerably developed and altered Maps and Survey which had come within my experience at the Royal Geographical Society since 1913.

For a third edition no such patchwork treatment was possible: the subject has changed too much. The present work has therefore been extensively altered and largely re-written. A brief chapter on early maps has been prefixed to the treatment of the modern, and a few pieces of typical maps have been reproduced to show the style of early periods. The maps analysed have mostly been published since the war. They are nearly all elaborately coloured; their merits could not be judged from small pieces, even if it were not too costly to reproduce them, and I have judged it no longer possible to give any idea of the range of modern maps from a few small specimens. It seemed best to enlarge the analysis and discard the illustration. I have made some attempt to deal with the new and difficult subject of Air Survey; the chapter on Geodesy has been entirely re-written; and the sections on Stereographic Survey much extended to cover important advances in this delightful subject. If these sections seem to be disproportionate in detail and in difficulty to the rest of the book, it is because the subject is new, and not as yet generally treated in textbooks. And if anyone should remark the frequency of reference for further information to papers in the *Geographical Journal* may I be allowed to say that the Society at its afternoon meetings for the discussion of technical subjects tries to deal adequately with all new methods, processes, and theories as they appear; and that since one may hope that a set of the *Journal* is within reach of all students of Geography, it seemed to be both natural and convenient to refer the reader thither for fuller treatment than is possible in a brief textbook.

I have to thank my friends Mr Edward Heawood, the Librarian, R.G.S., for much advice on the first chapter, and Mr Fawcett Allen, the Assistant Map Curator, for help in selecting maps for analysis. To the President and Council I am indebted for permission to reproduce many of the illustrations from the collections and the *Journal* of the Society, and I must thank also those who have allowed me to use their photographs of instruments.

Circumstances have much delayed the completion of this third edition: but the delay has made it possible to include many important developments of the last few years. I thank the Syndics of the Press for their forbearance.

A. R. H.

ROYSTON
July 1933

PREFACE TO THE FIFTH EDITION

THE Preface to the Fourth Edition, dated November 1941, explained that in the midst of a second World War it was not possible to give to the expiring Third Edition, dated July 1933, that thorough revision which the passage of years required. It was necessary to rewrite seven of the pages given to Ordnance Survey maps; other additions and corrections were made in a new chapter, at the end of the book.

This Fourth Edition has been consumed in eighteen months, during which a new series of One inch to the Mile maps of Great Britain with a kilometre grid has been placed on sale, and a new map on 1/625,000 has been published with a different metric grid. Long-distance flights have required new maps for studying great-circles and distances, and unfamiliar aspects of the world. These and other matters which seemed appropriate to the time are treated in a chapter of Further Additions to the preceding chapters.

I have as before to thank the President and Council of the Royal Geographical Society for permission to use material from the collections and the *Journal* of the Society.

A. R. H.

LONDON
January 1944

Chapter I

A BRIEF HISTORY OF EARLY MAPS

THE earliest maps which can be considered maps in the modern sense are the twenty-six which are included in the manuscripts of Ptolemy's *Geography*. The oldest surviving manuscripts date from the eleventh century: the maps are of two distinct types: and many scholars have argued that they are later constructions from the tables of positions given by Ptolemy. The best opinion seems now, however, to favour the view that they are descended from contemporary maps originally part of the work.

The Ptolemy maps were engraved in several editions before the end of the fifteenth century, the finest being the Rome editions of 1478 and 1490 with twenty-six maps engraved on copper. Modern maps were soon added to the original series, the Strasbourg edition of 1513 having twenty as a start.

The Portolan Charts of the Mediterranean and the Atlantic coasts form a quite distinct representation of geography. Their origin is much debated. The earliest dated examples surviving belong to the beginning of the fourteenth century, and the pattern then well established was copied with little modification, though with gradual extension, in manuscript charts and atlases of two centuries and more.

The portolan charts have the following characteristics: the outline of the coasts is very faint; the coasts are defined by the coast names written as thick as possible at regular intervals, inwards from the coast, the more important names in red; there is very little inland geography; islands and the deltas of rivers are brightly coloured or gilded; principal cities are drawn with towers and banners, gradually becoming more elaborate; the surface of the chart is covered with lines of loxodromes or compass bearings, radiating from points equally spaced on one or two circles, and later decorated with compass roses.

From these the manuscript World maps of the Catalan type were developed, with numerous large cities, banners, pavilions, and figures of emperors, animals, and occasional ships. The less the geography, the more numerous these decorative additions. A splendid example is the Este World map (*c.* 1400) preserved in the Biblioteca Estense at Modena; and the series culminates in the celebrated map of Fra Mauro (1459) in the Biblioteca Marciana—until lately in the Ducal Palace—at Venice, which embodies the geographical results of Marco Polo's travels, but in which the representation of the Mediterranean and Atlantic has finally lost its portolan character.

The globe of Martin Behaim at Nuremberg (1492) is the last monument of pre-Columbian geography, and the large manuscript map of Juan de la Cosa (1500) at Madrid in the Naval Museum the first map that has survived showing the discoveries of Columbus in the New World. A similar map by a Portuguese cartographer was obtained for the Duke of Ferrara by his envoy Cantino at the Court of Portugal, and is now in the Este Library at Modena. The first engraved map to show the New World is the small map by Contarini recently acquired by the British Museum, and dated a year earlier than the famous World map of Waldseemüller, a large wall map that was the first of a fine series that particularly distinguishes the sixteenth century, the most famous being the great World map of Mercator (1569), the first map constructed on his celebrated projection. The map on the same projection which Wright and Molyneux constructed for Hakluyt's *Principall Navigations* seems to be the first important World map made in England. Probably the finest of the whole series of large World maps is that of Hondius (1608) of which the only copy known is in the collection of the Royal Geographical Society, and has been reproduced recently in full size.

The atlases of the first half of the sixteenth century were successive editions of Ptolemy with more and more additional maps. The first really new atlas was that of Ortelius (1570), the *Theatrum Orbis Terrarum*, published in many editions and languages in following years. The first to use the name we now apply to such collections of maps was Mercator's *Atlas sive Cosmographicae Meditationes*... first published in its complete shape in 1595, though parts had appeared earlier in 1585 and 1590. The first general atlas published in England was that of John Speed in the edition of 1631, his first edition of 1611 relating to Great Britain only.

Plate I

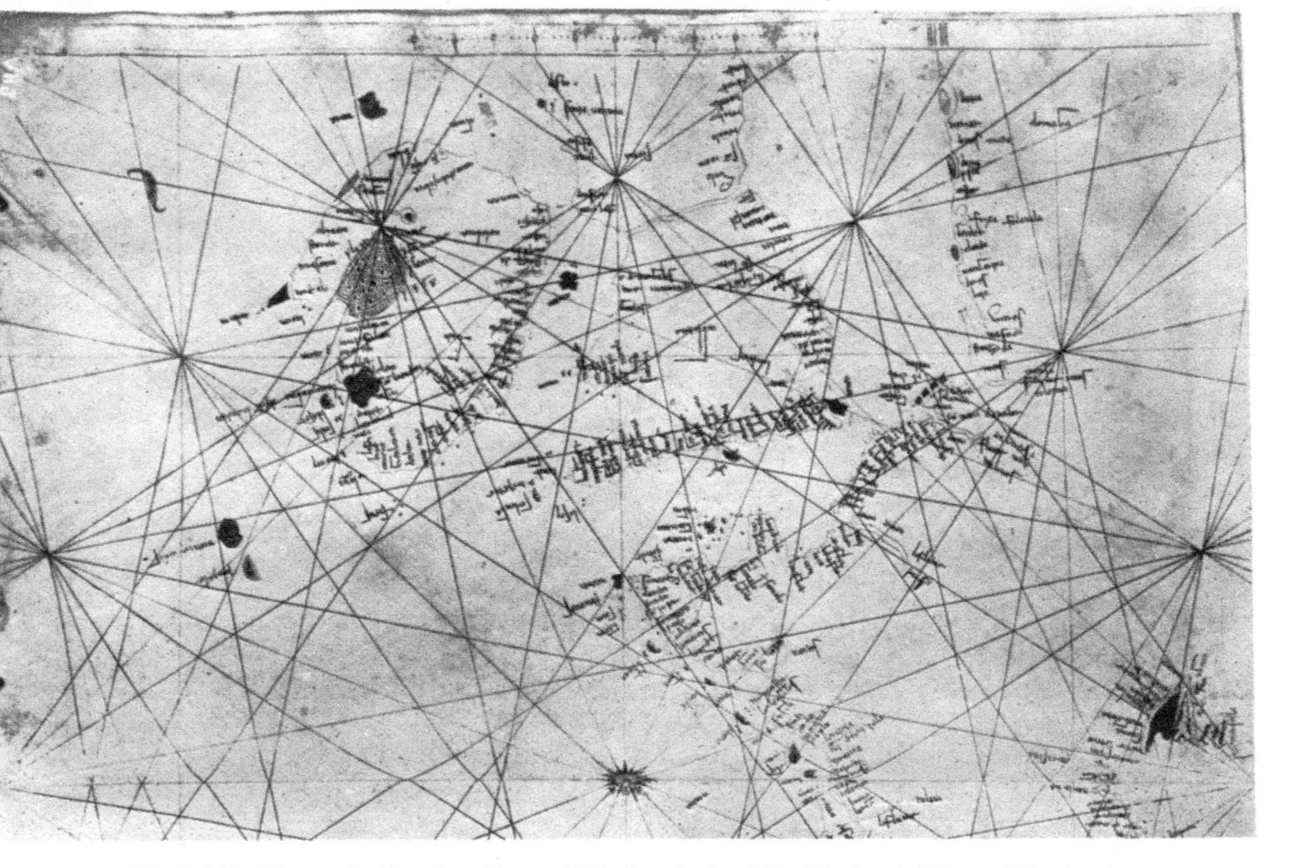

The British Isles on the Portolan Chart of Nicolaus de Combitis (Venice, Biblioteca Marciana).

Plate II

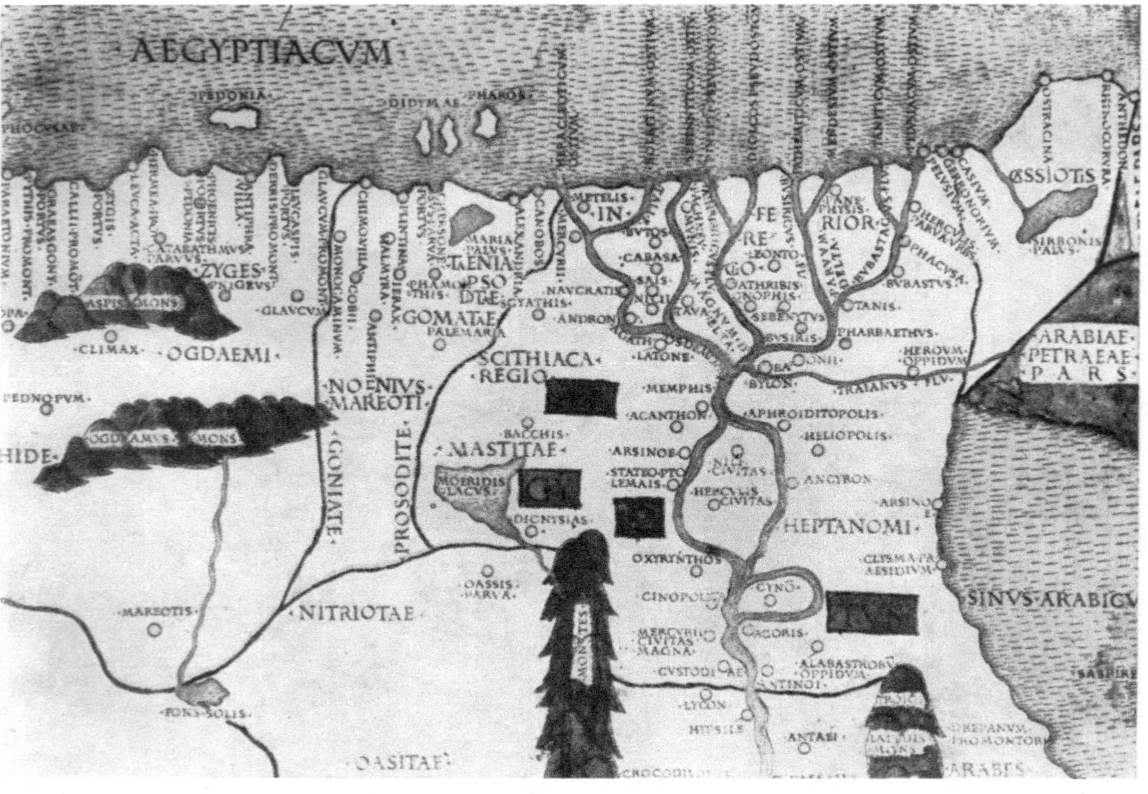

Egypt: from the Rome Ptolemy of 1490, plate engraved on copper for the first edition of 1478; names stamped.

Maps of the British Isles.

The British Isles were represented crudely on the earliest portolan charts that have survived: these maps are classified and discussed in a paper by Mr Michael Andrews (*Geog. Jour.*, December 1926). The British Isles are also shown in Ptolemy and in the Gough and Matthew Paris maps.

The earliest separate printed map of Great Britain is that of Pietro Coppo, recently found by Professor Almagia in the municipal library at Pirano in Istria, and described with a block in the *Geog. Jour.* LXIX, 441, May 1927. The first large map of England is that published by Lily in Rome in 1546, the source of his information being obscure. In 1564 Mercator published a very fine large map of the British Isles, of which no copy seems to have survived in this country, and only two in all. The first was found at Breslau in 1889 and was reproduced full size by the Berlin Geographical Society in 1892, with Mercator's large map of Europe and his World map. The material was sent to Mercator from England by a correspondent unidentified.

The most celebrated set of English county maps is that of Christopher Saxton, published between 1574 and 1579, and collected into a volume with a fine engraved title-page with portrait of Queen Elizabeth. Saxton also published a fine large general map of the country of which the first issue is unknown, though later editions are in the British Museum and the Royal Geographical Society. A little later John Norden began a description of England by counties, and completed Middlesex and Herts with maps (1593 and 1598) while maps of other counties made by him have been identified within the last few years. (See a paper by Mr Edward Heawood on "Some early County Maps", *Geog. Jour.*, October 1926.)

A select list of manuscript World maps.

The "Carte Pisane". Paris, Bibliothèque Nationale. 104 by 50 cm.

Undated, about 1300. Typical portolan chart with names in red and black, loxodromes in red and green, whose centres are arranged on two equal tangent circles, centred about Smyrna and west of Sardinia. Curious extensions by method of squares. No compass roses, drawings of cities, or banners.

The map of Petrus Vesconte of Genoa. Florence, Archivio di Stato.

Dated 1311. Shows Mediterranean eastward from Corsica; centre of loxodromes Eastern Aegean. No drawings of cities or banners, but effective chevron border, and curious crossed scales. Only inland geography a vague Danube. Islands and deltas coloured conspicuously.

This is the earliest dated portolan surviving, and the four next in date are by the same author, to whom are attributed also some curious maps in various manuscripts of Marino Sanudo's *Liber Secretorum*.

The map of Giovanni di Carignano. Florence, Archivio di Stato.

Undated, but first quarter of the fourteenth century, the work of a priest of Genoa. Important as not of portolan type, but for landsmen. Land coloured green, with names in red on white labels. All principal cities have large semi-heraldic town signs, not pictures except for a few outlying places. Caspian and Red Seas solid blue, the latter lined red; rivers blue. Handsome border. An original and interesting map. The reproduction in Nordenskiöld's *Periplus* very misleading.

The map of Angellino de Dalorto. Florence, Collection of Prince Corsini. 107 by 66 cm.

Dated 1325. Very fine condition and charming colour. Portolan type but some internal geography. Mountains green; rivers double green lines. Swiss lake. Prague in ring of mountains. Scandinavia a quadrilateral and triangle of mountains. Tigris, Euphrates, and Tower of Babel. Source of Nile south of Atlas. Many cities with banners. Probably the finest surviving map of this date. Reproduced in colour by R.G.S. 1929.

The Medici sea atlas. Florence, Biblioteca Laurenziana. 56 by 42 cm.

Undated, but from calendar probably 1351. Author unknown, but from internal evidence probably Genoese, though Genoa not named in the most important map, a planisphere with land coloured sepia, with names very small in white labels. Seems independent of Ptolemy, since Scotland not bent eastwards, and Nile quite different. Some figures of kings. Several other maps appear unfinished. Drawing good. Perhaps the most important early atlas.

The Catalan World map. Paris, Bibliothèque Nationale.

Dated 1375. Author unknown, but legends in Catalan. In beautiful condition, drawn on paper, mounted on boards to fold in sections like a screen. Centre of portolan type with fanciful extensions. Sea lined with wavy blue lines. Islands gaudily coloured. Great Britain treated as island, hence decorated and painted purple. Splendid figures and animals in outlying parts, with many castles in regular rows flying silver banners.

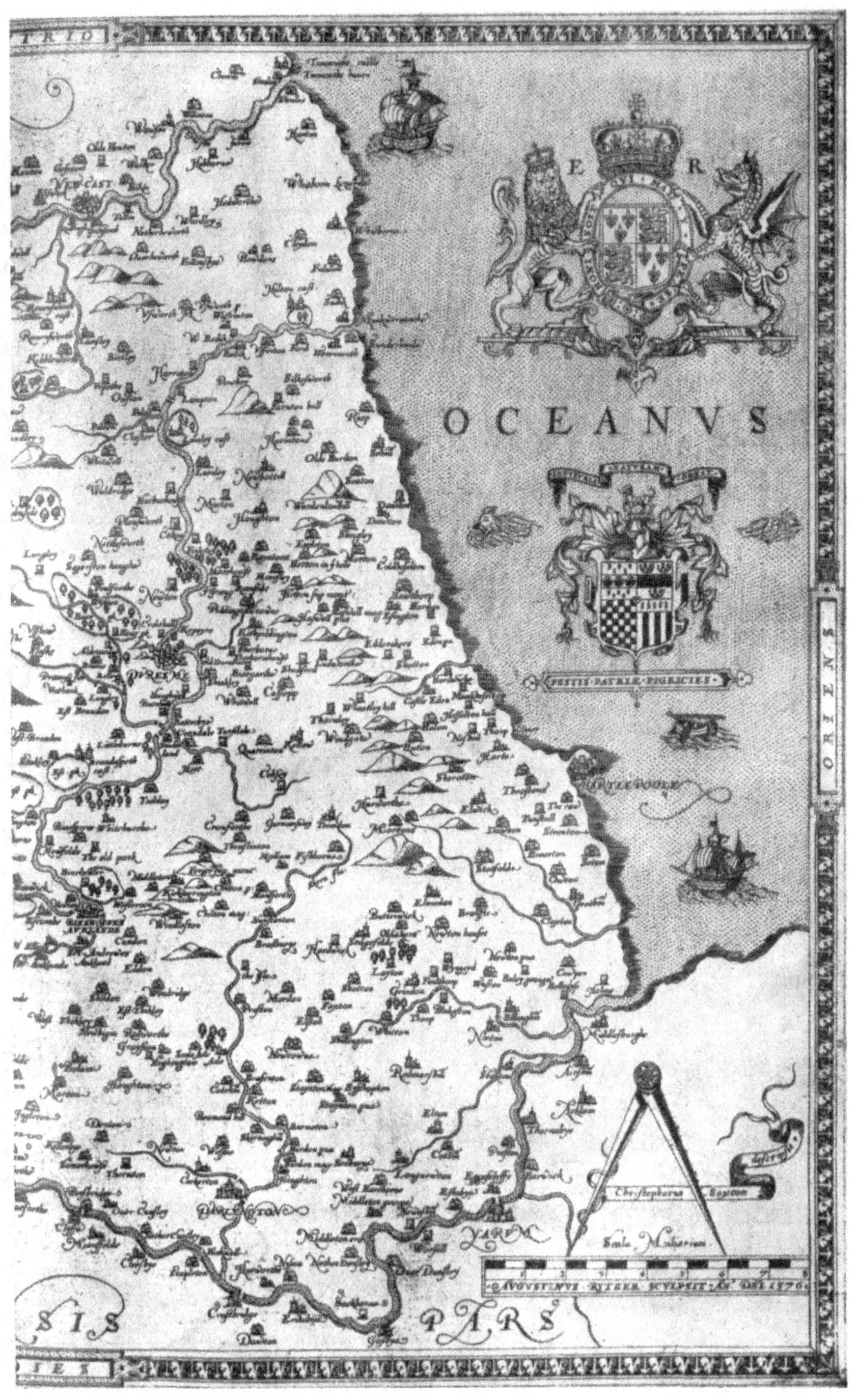

Part of Christopher Saxton's Map of the County of Durham, reduced to half scale. Engraved by Augustine Ryther 1576.

Plate IV

The Isle of Purbeck, from Speed's Atlas of 1611.

The Este World map. Modena, Biblioteca Estense. 115 cm. diameter.

Undated, but with detail (perhaps added) of fifteenth century. Circular on parchment in excellent condition. Legends mostly in Catalan. Has the same appearance of fantasy built round a portolan chart, but covers much greater extent than Paris map, showing Africa joined to great southern continent. Sea wavy blue lines, coast with coloured ribbon, rivers blue, mountains green. Cities conventional buildings with banners, kings in pavilions, mermaids, and a few ships. Colour of the whole map beautiful, and the most decorative of all early world maps.

Atlas of Nicolaus de Combitis. Venice, Biblioteca Marciana. 38 by 30 cm. (Plate I).

Undated, probably end of fourteenth century. The name is that of a former owner, not of the author. Typical atlas of portolan type with four maps: Black Sea and Aegean; Central Mediterranean; Britain, France and Spain; Portugal and Moroccan coast. The date 1368 assigned in the library catalogue arises from the legend I 368 (i.e. Insulae 368) attached to the usual mass of islands in a great bay on the west of Ireland: a curious error.

World map of Fra Mauro of Murano. Venice, Biblioteca Marciana. 1459.

This, the most magnificent of the surviving manuscript maps of the world, is a definite departure from the portolan type. There are few names on the coasts, and those not written perpendicular to it. The rivers are drawn in a heavy blue and white rope-like symbol. There are mountains with white tops, large numbers of great cities, especially in the outlying regions, a few ships at sea; but there are no figures, no pavilions, and no loxodrome lines. What would be empty spaces are filled with long descriptive legends in blue on the land and red on the wavy-lined blue sea. The principal names are in gold, and there is no black in the map. The general effect lacks contrast, and is not so decorative, nor the colour so good, as the Este map.

The Cantino World map. Modena, Biblioteca Estense. 220 by 100 cm.

Drawn in Portugal for Cantino, the Envoy of the Duke of Ferrara, in 1502, to show the discoveries of Columbus and Amerigo Vespucci. In fine preservation and well coloured, but wanting in spirit, and not to be compared with the Catalan World maps.

The above notes on some of the more important manuscript maps were made on a journey to study the question, Which are the maps most worthy of reproduction in colour by modern processes? Such reproductions as have been made hitherto have been printed from colour plates prepared largely by hand, and therefore of little use for critical study. The Royal Geographical Society is publishing a series of reproductions in facsimile, beginning with the map of Angellino de Dalorto from the collection of Prince Corsini, to be followed by the Este World map.

The principal manuscript maps surviving in England are of a different kind:

The Hereford World map of Richard de Haldingham. South transept of Hereford Cathedral. *c.* 1280.

A diagram rather than a map, built up round Jerusalem as the centre of the world; places more or less in correct relation to their neighbours, but entirely out of scale, and no attempt at correct outline even in the Mediterranean, where the portolans are so accurate.

The Matthew Paris maps of Great Britain.

Attributed to Matthew Paris, a monk of St Albans, and found in four manuscripts (three in British Museum, one at Corpus Christi College, Cambridge) of the monastery written about the middle of the XIIIth century. Based on Ptolemy.

The Gough map of the British Isles. Oxford, Bodleian Library.

From the writing the date is *c.* 1300. Quite distinct in style with little artistic merit, but raising interesting questions of nomenclature, and scarcely improved upon for two centuries.

A select list of early engraved maps.

Waldseemüller's World map. 1507.

Known in a single copy discovered by Father Josef Fischer, S.J., in the castle of Prince Francis of Waldeburg-Wolfegg at Wolfegg in Württemberg in 1901. Engraved on wood in eight sheets, based on the world map in the Ulm Ptolemy of 1486, and on the same projection. Longitudes much exaggerated so that there is room only for a narrow strip of the New World. Supposed until lately to be the first map with the name America. Reproduced full size with a learned commentary in *Die Weltkarten Waldseemüllers (Ilacomilus)* 1507 *und* 1516. J. Fischer und F. v. Wieser, Innsbruck, 1903.

Ortelius' World map. 1564.

In eight sheets, engraved on copper. Only two copies known, of which one is in the British Museum.

Mercator's World map. Duisburg. 1569.

Engraved on copper in eighteen sheets. Handsome decorated border but no pictures. Legends and figures on face of map, and comparatively few names. Not very artistic nor so well engraved as others of the period, but famous as the first map on Mercator's projection. Known from a single copy in the Stadtbibliothek, Breslau. Reproduced with the map of the British Isles and another by the Gesellschaft für Erdkunde, Berlin, in 1891.

Peter Plancius' World map. Amsterdam and Antwerp. 1592.

Engraved on copper in 18 sheets coloured and heightened with gold. Long known from the description by Blundeville of 1594 but supposed lost. Exhibited as anonymous by the Archbishop of Valencia at an exhibition in Madrid in 1893. Seen by Henry Hijmans from whose description Professor Van Ortroy of Ghent identified it in 1903. Examined by Dr F. C. Wieder in 1914 and published full size with commentary in his *Monumenta Cartographica* 1927, in 12 sheets. The original in the Collegio del Corpus Cristi at Valencia.

The World map of Hakluyt, Wright, and Molyneux. 1600.

Compiled on Mercator's Projection to illustrate the voyages of John Davis, and inserted in some copies of Hakluyt's *Principall Navigations*. Presumed to be the "New Map with the Augmentation of the Indies" mentioned by Shakespeare in *Twelfth Night*, iii. 2. Reproduced by the Hakluyt Society in 1880 with a memoir by Mr Coote.

Father Ricci's World map. Peking. *c.* 1602.

Painted on a screen now in the Library of the former Imperial Palace (of which a photograph of part is reproduced in *Geog. Jour.* LXIX, 532, June 1927) and engraved on wood blocks to make a map 12 feet by 6 feet. Seemingly based on Ortelius for the projection and Plancius for the content, rearranged to bring China into the middle of the map. The example, perhaps unique, shown in the map room of the R.G.S. is a later issue since it bears the ideographs for "The Great Manchu Kingdom" across China. Described by Dr Lionel Giles in *Geog. Jour.* LII, 367 and LIII, 19, Dec. 1918 and Jan. 1919.

The World map of Jodocus Hondius. Amsterdam. 1608.

Engraved on copper in 12 sheets 18½ inches by 13½ inches with borders made up from smaller sheets.

This, the most splendid example of the Dutch school of engraving, is known only from the copy acquired by the R.G.S. in 1919, and now shown in the ambulatory of the Society's House. In 1926 it was reproduced to full scale in collotype, with a memoir by Mr Edward Heawood and translation of the many interesting legends engraved on the face of the map. The geography is in the main that of the Hondius atlas of 1606, the style similar to the World map of Blaeu of 1605, which may, however, have owed much to Hondius' earlier large World map of about 1602 now lost, while both Hondius and Blaeu were inspired by Plancius. The lettering deserves study as the finest of the time, perhaps of any time.

The Hondius World map of 1611 or later. Amsterdam.

The only known copy was discovered by Father Fischer, S.J., in the collection at Wolfegg Castle. There are reasons for believing it first published in 1611 but the only copy shows Le Maire strait discovered only in 1616. Resembles Blaeu of 1605 more than Hondius of 1608, and perhaps published in hemispheres on the stereographic projection to compete with that map. But the border original, with pictures of animals instead of peoples. Remarkable as showing the magnetic variation at a number of places, and ocean currents. Reproduced to full scale in 1907 with memoir by Professor Luther Stevenson and Father Fischer under the joint auspices of the American Geographical Society and the Hispanic Society of America.

Several of the most important engraved maps of the above list survive only in a single copy, and it is known that important maps by the same geographers were published but are lost. Such are Mercator's map of America in fifteen sheets whose sale is recorded in the Plantin Press ledgers from 1581, but of which only the engraved cover and title survive in a single copy found recently in Italy ("Lost Mercator Maps", *Geog. Jour.* LXII, 138, Aug. 1923); the English edition of Peter Plancius' map published in London in 1595; and a large World map by Hondius on Mercator's projection published about 1602. These large maps have been lost because they were generally mounted on canvas for the wall; they were too large to be framed conveniently; they have worn out, and been thrown away. The few survivors are more often than not the separate sheets unmounted or bound in collections.

The student of early maps is often called upon to decide whether a sheet is of the original issue or a later, perhaps much later, impression from the old plate. He must therefore study the watermark of the paper. A valuable introduction to this subject, with many illustrations, was published by Mr Edward Heawood in the *Geographical Journal* for May 1924. Hondius engraved a map of the world in hemispheres showing the routes of Drake and Cavendish, which might well have been made before he left England, but the watermark is that of paper known only from Amsterdam and Western Germany. The same paper is used for an undated map of France by Hondius, of which the R.G.S. has the only copies of three of the four sheets, the last being unknown. The date may be taken as about 1600, for the paper was used at Cologne in that year, but not by Hondius in any later-dated atlas. These are simple examples of the use of watermarks.

The paper of Speed's first edition of his *Theatre of the Empire of*

Great Britain is that used by Hondius for several of his editions of Mercator; the work was therefore probably printed, as well as largely engraved, at Amsterdam.

Globes.

The new aspects of the world revealed by the discoveries of Columbus and of Magellan were more conveniently represented on globes than on maps. We find accordingly great activity of early sixteenth-century engravers and instrument-makers in constructing globes for the libraries and cabinets of princes and cardinals. The principal authority on the history of globes is Professor Matteo Fiorini, on whose works Professor Luther Stevenson based the most complete treatise in English, *Terrestrial and Celestial Globes, their History and Construction*..., in two well-illustrated volumes published for the Hispanic Society of America in 1921.

Of globes painted on wood we may mention Martin Behaim's preserved at Nürnberg; the pair made for Pope Julius II about 1500, of which only the celestial survives in the Vatican Library, 95 cm. in diameter; and the Green Globe of *c.* 1515 in the Bibliothèque Nationale. Of globes of metal there is the Laon Globe contemporary with Behaim's and pre-Columbian in its geography, though it gives the date for a Portuguese discovery in 1493. The Lenox Globe in the New York Public Library of copper gilt is the earliest metal globe to show the New World. It may be dated about 1510. A beautiful globe engraved on silver after Gastaldi, about 1536, is deposited by Mr R. S. Whipple in the Museum of the R.G.S. It is not known who first thought of engraving the World map in gores which, when wet, could with a little manipulation be stretched and squeezed on to the surface of a globe. But there is evidence that Waldseemüller in 1507 was engraving gores at the same time as his World map, and a set of twelve in the library of the Prince of Liechtenstein may be the only surviving example. The earliest globes made up from engraved gores seem to be those of Johann Schöner of 1515 preserved at Weimar and Frankfurt, and from that date there are many makers.

Mercator published his first terrestrial globe, of 41 cm. diameter, in 1541 and a number survive in gores or mounted. Blundeville speaks of Mercator's globes as common in England at the date when Emery Molyneux, at the expense of William Sanderson, caused Hondius to engrave the famous globes about 26 inches in diameter, of which the only survivors are the pair in the library of the Middle Temple, unless an undescribed pair at Cassel prove to be the same. Since it is almost impossible to photograph a varnished globe, students may hope that a set of the gores may be found some day and reproduced.

Chapter II

THE MODERN MAP

TOPOGRAPHY means, in the original Greek, the description of a place. A map is a drawing to illustrate such a description. By gradual steps it has been found possible to make these drawings in such a way that they convey the necessary information without the description in words, and what is now called a topographical map can give, to those who know how to read it, all the information which is wanted to enable a man to judge of the natural obstacles which may hinder his getting about the country, or the means of communication which have been established by man to facilitate his journey. In this book we shall be concerned chiefly with these topographical maps, but shall have something to say also on cadastral maps, atlas maps, wall maps, and nautical charts.

The amount of information which can be compressed into a map depends in the first place upon the perfection of the system of conventional signs in which the map is drawn and skill of the draughtsman; secondly upon the size of the map in comparison with the ground which it represents, or, in other words, upon the scale of the map; and thirdly upon the perfection of eyesight presumed in the user. There is something to be said for the presumption that the user may be supposed to have a magnifying glass. It is clear that the larger the scale, the more detailed is the information which can be given. But, on the other hand, the larger the scale, the greater is the number of sheets required to cover a given area, and the more cumbrous is the map to use. For purposes of local administration and government it is necessary to have maps on which every separate property, however small, is distinctly shown; but such detail would be only confusing to those who wished to learn the way about the country. And again, the details of roads, the shapes of the hills, and all the other information essential to the traveller on foot or on wheels, are equally unessential to those who use the map for the study of broad questions of history, of commerce, or any such matters.

Hence we shall find that maps may be classed in three principal divisions:

Cadastral maps, on large scales, show boundaries of property, and individual buildings. They are required for local administration and taxation; the management of estates; the identification of property in legal documents; and for detailed affairs of every kind.

But it should be remarked that the English map on the scale of 1 in 2500, commonly called the 25-inch map, shows the visible hedges and fences, whereas the real boundary of the property is very frequently some feet beyond the hedge. Hence the 25-inch map is not strictly a cadastral map, though it is commonly called so. Moreover, a cadastre proper records the co-ordinates of boundary stones, especially corners, referred to the general projection or framework of the survey. There is nothing like this in the British system. On the other hand, the cadastral maps in other countries are usually in manuscript only, deposited with the local authorities and Great Britain is the only country which publishes on a scale anything like 1/2500. The registration of co-ordinates is more important in a country which has no permanent hedges or boundary walls, especially in Egypt, where the annual inundation of the Nile is apt to obliterate the very vague boundaries of cultivated lands.

Topographical maps show the natural features of the country, hills and rivers, forests and swamps; and in addition such artificial features as man has added to the country in the shape of towns and villages; roads, railways, and canals; bridges and telegraphs. They serve as guides for travel on business or pleasure, or for the operations of war. They are on smaller scales than the cadastral maps, and cannot show the boundaries of individual properties. The inch-to-the-mile map is the standard topographical map of the British Isles.

Atlas maps are on scales still smaller. Most of the topographical details have been suppressed; only the principal ranges of hills and the main streams of the rivers, the chief towns, and perhaps the main lines of the railways can be represented, for they aim at representing on a single sheet a whole country, a continent, or even the World.

Nautical charts differ from maps in that they leave the land blank except for detail within sight of the sea. They have a special system of conventional signs to show buoys, lights, and other navigation marks; they show submarine contours only within soundings; their depths are referred to low water as datum instead of to mean sea level; and they give information on tides and currents.

Wall maps are generalised and boldly drawn so that they can be read by a whole room, but otherwise are like atlas maps. It is important to note that maps to be made into lantern slides should usually be drawn in the style of wall maps. Others are generally too crowded to make effective slides for projection.

The necessity for maps.

It is a little hard for one who lives in a country long settled and completely mapped to realise the difficulties which confront at every turn those who find themselves citizens of a new and unmapped country. The need of maps appears most urgently in case of war, and in the past it has usually happened that military necessities first compelled the production of a map. The beginning of the Ordnance Survey of Great Britain may be traced, for example, to the Highland Rebellion of 1745. In peace the necessity of maps for all purposes of administration is less conspicuous, but none the less real. Until a country is mapped it is impossible to devise any well-considered schemes of communication; until it is surveyed on a fairly large scale it is impossible to make grants of land to settlers, or at least it is impossible to give them a clear and undisputed title to their holdings. In larger affairs the need of maps is equally great. Until a country is mapped it is impossible to agree upon a boundary which shall be satisfactory to both sides, and in the absence of maps the most costly mistakes have been made through sheer ignorance.

Again, until a country is mapped there is continual waste in making surveys for special purposes. If a railway is projected it is necessary to make a special survey of the proposed course of the line. This survey costs a large sum of money, but since it is made for a special purpose it is not complete. It generally has good levels but bad azimuth, sometimes depending on magnetic compass only, and rarely with any accurate determination of latitude and longitude; it rarely pays any attention to the topography of the country on either side of the line, though one would have thought that access to the railway was most important. The railway survey is never published, and cannot be used as a contribution to the proper mapping of the country. In the earlier part of the nineteenth century, while the Ordnance Survey was in the initial stages, several million pounds were spent in England on railway surveys, surveys for the commutation of tithe, enclosure of common lands, and so forth, nearly all of which money might have been saved had the regular survey of the country been finished earlier. It is the truest economy to push forward the survey of a country at the earliest possible moment.

Very many expensive mistakes have been made in the past because responsible officers of Government have not been trained to discriminate between maps which are reliable and those which are not. Thus the Treaty of Paris of 1783, which defined the boundary between Canada and the United States, was based on the geography of the very imperfect Mitchell map of 1755. The assumption that the Lake of the Woods had a recognisable north-west corner, and that the parallel drawn west through that corner intersected the Mississippi led to difficulties which were not settled for nearly a hundred years. More recently the belief, based on Stanley's necessarily imperfect longitudes, that Ruwenzori and the Mufumbiro Mountains lay east of the 30th meridian caused great difficulties in the delimitation and demarcation of the Uganda-Congo boundary. The assumption, in the absence of any surveys, that the Atlantic-Pacific watershed coincided with the crest of the Andes was the cause of the Argentine-Chile boundary dispute. The assumption that a watershed will naturally be a well-marked feature is often falsified, as on a small scale in the new Austro-Italian boundary at the Brenner Pass, and probably on a much larger scale when the Labrador boundary comes to be demarcated on a recently glaciated plateau where the original drainage has been confused by moraines.

It is a good principle that no agreement for an international boundary, or indeed any division of land, should be made until that land is mapped. But it should be noted that in treaties we usually find the stipulation that if the text and the accompanying maps disagree, the text shall prevail.

Conventional signs.

In order that it may be possible to compress as much information as is required into the minimum of space, to ensure clearness and legibility, it is necessary to adopt a carefully considered scheme of conventions, so that the character of every line and the style of every letter may convey a definite meaning. The map thus becomes a kind of shorthand script, whose full meaning cannot be understood without a thorough knowledge of the system employed. It is therefore clearly advantageous that map-makers should come as soon as possible to some general understanding on the subject of conventional signs, in order that it may not be necessary to learn, as it were, a new language whenever one tries to read a new map.

The resolutions of the International Map Committee, which met in London in 1909, and again in Paris in 1913, to frame a scheme for the one-in-a-million map of the World, are in this respect of great importance, and we shall often have to refer to them in what follows. But it is important to remember that a scheme which is suited to a map to be printed in many colours may be quite unsuitable for a map with few colours or in one only. Moreover, too rigid adherence to a convention is fatal to any attempt to improve the artistic appearance of the map. Again, conventional signs which are suitable to a finely engraved map are often too difficult to draw in the field, and must be replaced by others.

The characteristic sheet.

The characteristic sheet is the key to the system of conventional signs employed on the map. Usually, though not invariably, each sheet of a map bears a small characteristic sheet of the principal conventional signs; but for a complete understanding of the system it is always necessary to refer to the full characteristic sheet that is published separately.

The "reference" is usually a small characteristic sheet of the principal conventional signs, with statement of the vertical interval of the contours, the unit of length for heights, and similar information, which should always include the date of the survey and of its revisions, the date of publication, and indication of the sources from which the map is derived, if it is not the product of a uniform survey, but is compiled.

Each sheet of a series should show the names or numbers of the adjacent sheets, and it is convenient that these should be shown on a little diagram with an outline of the principal features, to serve as a ready guide to the extent of the sheet. In a series such as that of the International Map it would also be convenient to have a small index to boundaries, such as was drawn on the provisional sheets of that series compiled by the R.G.S. and issued by the G.S.G.S. during the Great War.

Roads.

In reading the representation of roads on a map, it is necessary to remember first of all that the road is not represented true to scale in width, but that the width is conventional, signifying the class of the road. Such a convention is clearly necessary. On the scale of one inch to the mile a road sixteen yards broad would, if shown true to scale, be only one-hundredth of an inch broad; and this is far too little for legibility, or to allow distinctions between the representation of different classes of roads. Hence roads must be shown of a conventional width; and when the scale of the map is doubled it is by no means necessary that the road shall be drawn twice as broad as before.

The breadth of the road symbol necessarily obscures the smaller sinuosities of the road, and especially the characteristic shapes of road junctions and crossings. This is the more inconvenient in towns, in whose plans the road symbol might with advantage be narrowed.

A distinction should be made between fenced and unfenced roads, the former being shown by double continuous lines, the latter by dotted lines. The distinction is of military importance, since fenced roads afford cover and obstruct the free passage of troops across country. Grades of metalled roads may be distinguished by differences in breadth, and the first two grades coloured. Unmetalled roads or cart tracks may be shown by still narrower double lines, and footpaths long-dotted. But it is generally a mistake to colour as first-class roads the residential roads in the suburbs of a city which, though maintained in first-class condition, are not main roads. They obscure the direct routes through a town, which are difficult enough to follow on the map without this complication.

The conventional signs for roads on the Ordnance Survey maps were elaborated after the conference between various road authorities and users held in 1911–12. The main roads between towns were then coloured a strong red, secondary roads are yellow—both between the black lines of the old engraved map. They thus became almost too prominent. The newly drawn map of Scotland preserved this type of road symbol, so that the disproportionate strength of the roads on British maps could no longer be explained as due to the effect of adding colour to an existing engraved black plate. In the new one-inch map of England this disproportion is remedied. The road is brown instead of red and the general tone of the map is buff. Roads shown in colour without black outlines are not sharp enough, and are suitable only for small-scale maps.

In general it may be said that all existing topographical maps fail in the representation of hill paths. These have an importance quite different from paths of equal quality on the level. They are very generally well-marked and permanent features of the ground, and should be shown much more conspicuously than they are. On most maps they are very hard to follow when they run through hill-shaded and wooded ground. It seems reasonable to demand that all well-made and easily followed paths in mountainous parts of Great Britain should be shown by a single line of orange and black.

The want of some such convention is particularly marked in holiday country. For example, in the Lake District there are many paths well made, and as permanent in location, and very nearly as obvious to follow, as the high roads. Such are the principal ways up Helvellyn, or the path over Grisedale Pass from Patterdale to Grasmere. On the half-inch map with layers these are not shown. Again, in Switzerland the principal paths are well made and most elaborately marked with signs and patches of colour on the rocks. But it is almost impossible to follow these paths on the 1/50,000 map. It would seem that from the military point of view also it must be of importance to show these paths.

Railways.

These are almost universally shown in black. There is generally a distinction between double or multiple and single tracks, narrow-gauge railways, and tramways. The most recent continental maps have much elaborated the signs for railways, distinguishing those worked by electricity, those with rack on the steeper sections, as in many Swiss railways, and purely mountain railways, all rack; and it is clear that the distinction is important, especially from the point of view of public safety on the one hand, and of military adaptability on the other. When gauges differ, it is often desirable to write the gauge at intervals along the line.

It is often useful to have special railway maps, in which the other topographical features are subordinated. The Swiss railway maps are excellent examples of the best system for this purpose. The general map is printed in light brown, and the railways, with names of the stations, are overprinted in heavy black. This is far better than suppressing the topographical detail, as is too often done in railway maps.

The representation of railways running underground through towns involves difficulties which have not been successfully overcome.

Rivers.

Rivers and streams should be shown always in blue. Their importance is great, not only the obvious importance which they possess as sources of water supply, means of communication, or obstacles to getting across the country, but the subsidiary yet very real importance which they derive from their use in helping realisation of the relief of the land. When the streams are shown conspicuously in blue they make it easy to follow the run of the valley bottoms, and to distinguish valleys from ridges. But difficulties arise if the streams are shown in the heavy cobalt blue that brings up the relief; where they broaden into lakes, or open into estuaries, they suddenly become inconspicuous. The solid blue must not stop abruptly, but be shaded off into the estuary, and this is helped by blue waterlining along the shore. But the difficulty cannot be entirely overcome, or has not been, and there is room for experiment.

There should be special signs for locks, weirs, and falls, and indication of navigability of rivers. There should be a distinction between natural streams, canalised streams, and wholly artificial canals, though confusion is avoided by the consideration that the shape of natural and artificial waterways is very different, and that the latter tend always to run along the contours.

A further difficulty is sometimes caused by the failure to show streams that run in deep ditches by the roadside, which is not uncommon in flat country such as the fen districts of England. It should be easily possible to show the necessary blue line alongside the black line which borders the road; its absence makes it impossible to discover what becomes of a stream which runs by the roadside.

In general a stream will cut a contour at right angles to its general direction, the contour being thrown back upstream where it crosses. When the fall of the stream is rapid, and particularly when the ground is rocky, the contours are V-shaped at the crossing; when the ground is flat and alluvial the crossings of the contours are more rounded. It is clear that in general a stream cannot cross a contour more than once. There are, however, exceptions to this rule which are at first sight puzzling. They occur in very flat country, where the streams are sometimes confined by banks much above the level of the ground on each side. In such cases it is possible for a stream to cross a contour several times. It would be well that streams of this character should be distinguished by some special sign, for they are important from the fact that it is easy to breach their banks and flood the surrounding country.

Woods and forests.

In black engraved maps these are often shown by small tree signs, sometimes of two varieties to distinguish between deciduous trees and conifers which are mostly evergreen. The effect is apt to be heavy, obscuring details, and rendering names almost illegible. On coloured maps these tree signs may be on the green plate, but this is rarely successful, and it is often better to overprint black tree signs with a solid tint of pale green. But on layer-coloured maps this solid green interferes with the green layers, while tree signs in green are hardly conspicuous enough on the green layers. Much depends also on the size of the woods, and on the steepness of the ground. Extensive forests in mountain country can hardly be represented in green without ruining altogether the effects of layer colouring. So much depends on the other contents of the map that one cannot lay down any rules for the representation of woods and forests, except that where tree symbols are used they should be disposed irregularly to represent woods, and in rows to denote orchards.

The representation of hill features.

The real difficulty in map making is to represent the relief of the ground. This is naturally so, for we are trying to find a method of representing a solid figure by drawing on a flat sheet, or to represent three dimensions in two. Moreover, we have not a free hand to do the best we can with this particular problem, but are limited by the condition that we must not obscure the other details on the map.

Until recently little progress was made in the solution of this problem. So long as the map was printed in a single colour, almost necessarily black, very little could be done. But improvements in the processes of colour printing made it possible to produce maps in colour, sometimes with twelve or fifteen separate printings. This altered completely the conditions of the problem, and great progress had been made, when the enormous increase in costs arising out of the war made a very large number of printings impossibly expensive. The ingenuity of map-makers has therefore been turned to the problem of producing the best effects with a limited number of printings.

The relief of the ground may be shown in a number of different ways: by hachures or hill-shading; by contours; by spot heights; by hypsometric tints, generally called in England "the layer system". And these methods may be, and often are, combined so that three or four are in use on the same map. We shall begin by considering them separately, and then discuss how they may be used in combination.

Hachures.

Hachures are lines drawn down the directions of steepest slope. On slight inclines they may be delicate and not too close together. As the slope gets steeper they may be drawn heavier, and be closer together. If they are drawn faithfully they are very expressive, and give a good idea of the shape of the ground. But the system of hachuring has the following defects:

Its range is small; that is to say, it is impossible to show many different degrees of slope. Slight folds in the ground, if they are shown at all, are exaggerated; really steep slopes cannot be shown proportionately heavy without obscuring all other detail. Moreover, it is impossible to preserve uniformity of treatment on the different sheets of one and the same map. The steepest slope on a sheet of nearly flat country may be actually less steep than a relatively moderate slope on a mountainous sheet; and slopes which are important in the former may be insignificant in the latter. Hence hachures can be used to indicate that there is a slope, but they cannot give much information as to the absolute degree of the slope.

It is difficult to draw hachures properly in the field; and when the field sheets come into the hands of the engraver it is certain that they will be improved in appearance, and generalised, so that they become untrustworthy in detail.

Hachures, then, belonged to the days of delicate and expensive engraving upon copper, and with the changing conditions of map reproduction they are rapidly becoming obsolete. Excellent examples of the system are to be seen in the old engraved sheets of the one-inch Ordnance Survey maps of the United Kingdom, and in the older maps of Switzerland; and it survives on a few modern colour-printed maps that have as their basis transfers from the engraved plates; see for example the new one-inch O.S. map of England, in which transfers from the old hachures have been used with great effect.

The term hachure has the definite meaning assigned to it here, and should not be confused with hill-shading, described in the following paragraph. It is regrettable that the distinction between hachures and hill-shading has not always been maintained in manuals of map reading.

Hill-shading.

Hill-shading aims at producing very much the same effect as hachures in an easier and cheaper way. The draughtsman colours the slopes in rough proportion to their intensity, with brush or stump. The lithographic draughtsman copies this drawing upon a grained stone; or the drawing is photographed and a hill-shading plate produced in "half-tone", by a process similar to that by which photographic illustrations are produced in books and newspapers. Thus hill-shading consists of series of dots, whereas hachuring consists of series of lines; and the two are very easily distinguished.

It is difficult to do hill-shading effectively in the field, and it is generally added by the draughtsman in the office; but since it is in its nature more generalised than hachuring, and is quite unsuitable for showing detail, this is of small consequence.

Both hill-shading and hachuring suffer from the defect that they do not readily show which direction is uphill and which is downhill. By themselves, in fact, they cannot give any indication at all on this important point. A ridge and a valley, each with shaded sides, cannot be distinguished one from the other simply by the shading, and one must rely on other details, such as rivers in the valley bottoms, or heights marked at various places, to indicate the interpretation

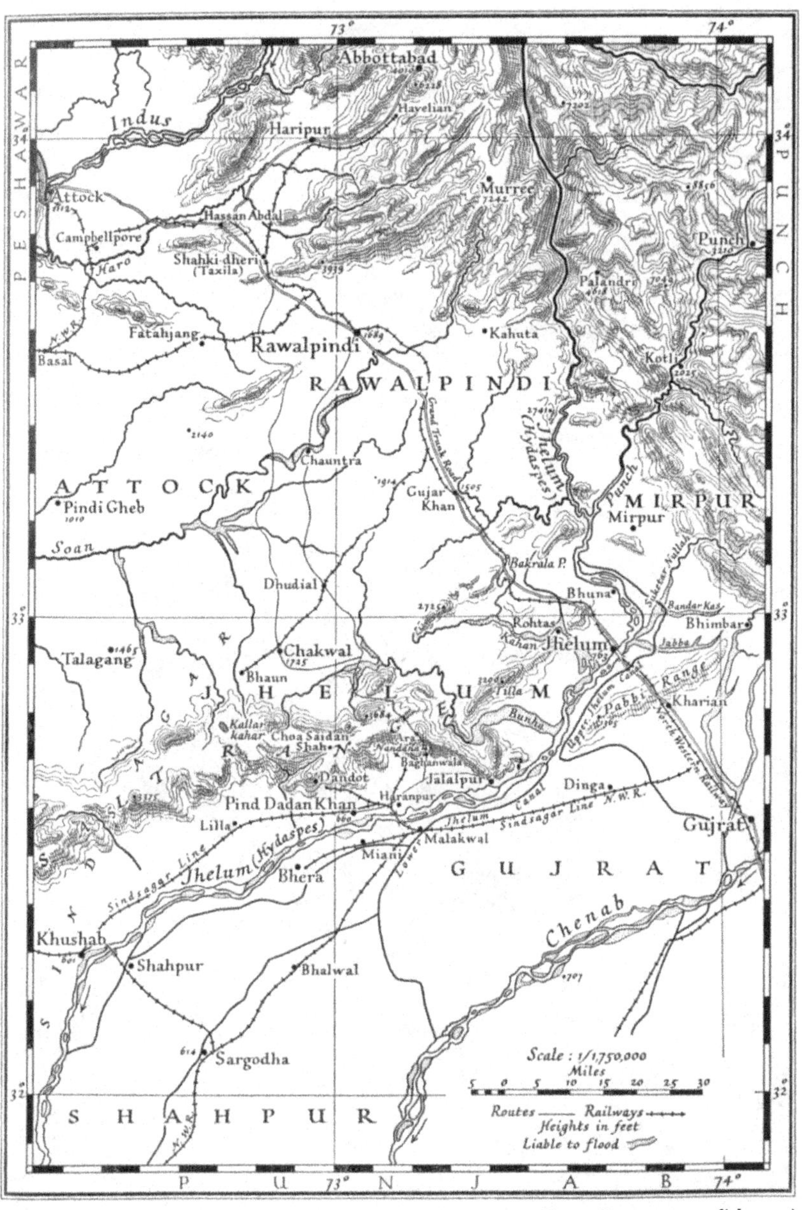

(*Geogr. Jour.* LXXX, 32, *July* 1932)

Quill-written Roman lettering and hill-shading by cross-hatching contours.

Plate VI

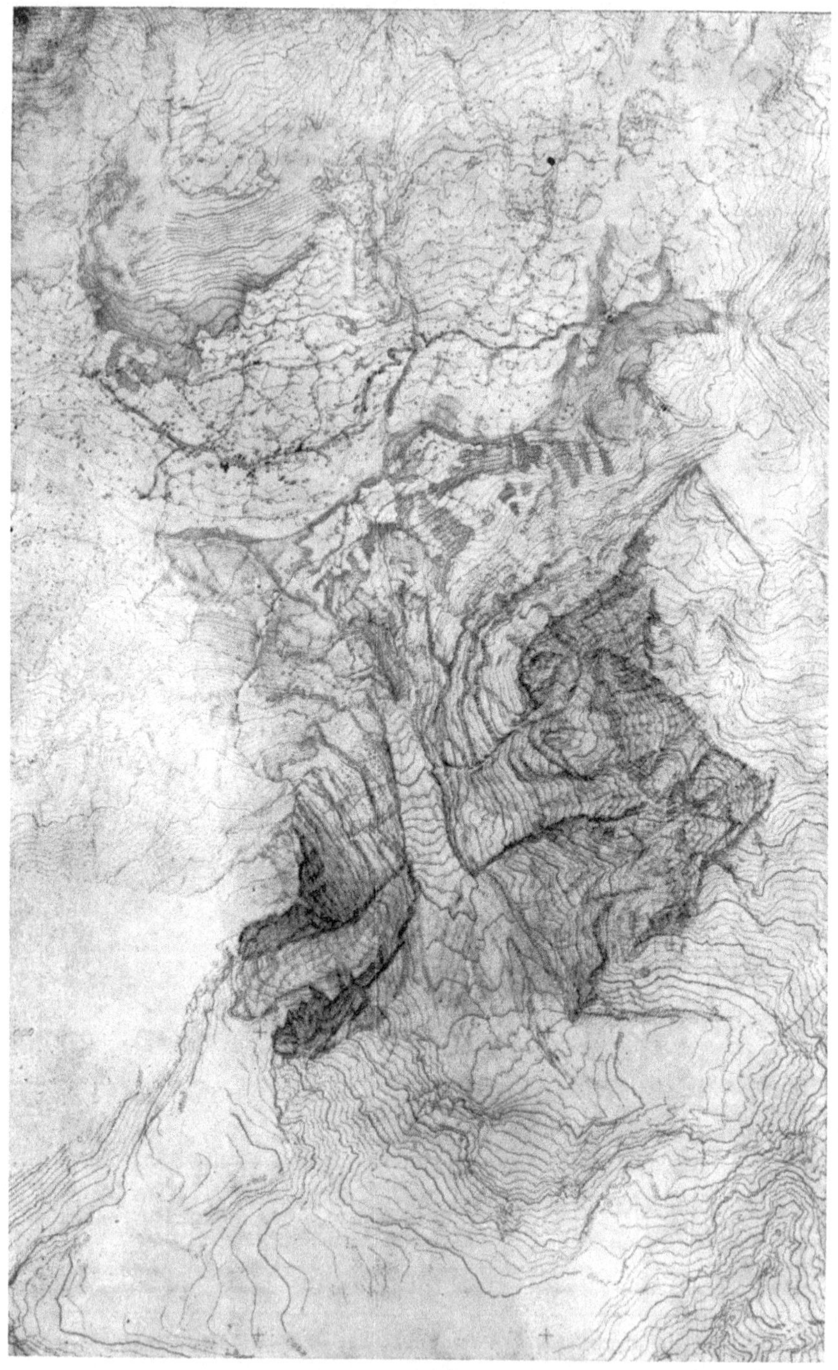

(*From paper by Mr Michael Spender: "The new Photographic Survey of Switzerland", reproduced by permission of the Director, Swiss Federal Surveys. Geogr. Jour.* LXXIX, 383, *May* 1932)

Contour plate plotted on the Autograph worked up in pencil by topographer in the field to improve the representation of rock faces.

of the shading. A sheet with hill-shading and nothing else might be interpreted equally well in opposite ways.

We are speaking now of shade which depends only on the degree of slope, without any convention as to the way in which the shade is cast; without, indeed, any possibility of imagining it as a shade produced by contrast with light from a definite source. It is sometimes spoken of as vertical shade—an expression which can have none but a conventional meaning; in other places it is called the shade cast by a vertical light, which is meaningless. These expressions serve, however, to distinguish it from the other form of hill-shading, which represents the shadow which would be cast on the ground by oblique light from a low source, such as the sun near setting. For some reason which is not clear the source of light is generally imagined in the north-west. Oblique hill-shading is very much used in European maps, and it produces a strong effect of relief. But it has the disadvantage that it makes the slope in the shadow look steeper than the slope in the light, left unshaded. In fact it is only by induction that one gets any suggestion at all of the slopes towards the light.

We may compare thus the effects of the vertical and the oblique systems of shading. Imagine a ridge and a valley, each running north-east, with the slopes on each side the same. Under the vertical shade they would be indistinguishable. Under the oblique shade they would be readily distinguished one from the other; but the ridge would appear steeper on its south-east side, the valley on its north-west side.

To avoid this difficulty an ingenious system was devised for a new map of France on the scale of 1/50,000. The map was hill-shaded on both systems, printed in different colours. There was a vertical shade in bistre, and overprinted was an oblique shade in purple-grey. On moderately undulating ground the effect is excellent; in mountainous country the colour becomes too heavy. The system is of course expensive, and has for the present been abandoned on the French maps, but has just been applied to the hachures on the new O.S. one-inch map of England.

It is clear that no system of hill-shading can give any information as to the actual height of the ground above sea level. Hill-shading should therefore be considered as a means of bringing up the relief of the country graphically, to supplement the more precise information which can be given in other ways. This being so, it is important to notice that the hill-shading should be kept very light. The least tint that will serve to show the difference between one side of a hill and the other is enough, since the degree of slope should always be denoted by the closeness of the contours. Hill-shading often suffers from being too heavy. A pale transparent grey is probably the best colour.

Spot heights.

Spot heights are the heights above sea level marked at various points on the map. They are sometimes called spot levels; but the misuse of the word level in this connection is to be condemned.

On cadastral maps the height of each bench mark is usually given; these are not strictly speaking spot heights, since they refer to the height of the bench mark, not of the ground. The heights which, on the British cadastral maps, are given along the crown of the road are spot heights, and a selection of them is given on the smaller scale topographical maps. There is a strong tendency to give spot heights for summits, and not for the bottoms of depressions.

Spot heights are very useful as exact points of reference, but they cannot by themselves give much idea of the form of the country.

Contours and form lines.

A contour is a line joining a series of points which are all at the same height above mean sea level. If the sea suddenly rose a hundred feet the new coast line would follow the hundred-foot contour.

A form line is an approximate contour, not accurately surveyed, but sketched upon the ground.

There is some difference of practice in the use of the two expressions, but for our present purpose it is not material to define the precise difference in degree of accuracy which divides contours from form lines. It will be sufficient to keep the term contour for the product of instrumental methods, even if rough, and to call the much less accurate contours which are merely sketched, form lines.

Contours give a maximum of precise information with a minimum of obstruction to the map. They may be considered the standard method of showing relief, to which all others are subsidiary. But to be effective they must be drawn on carefully considered principles.

Contours should be drawn at uniform intervals of height; any departure from this rule is certain to lead to inconvenience. The interval chosen will naturally be a simple multiple of the unit of length employed: 50 or 100 feet; 10 or 20 metres.

The contours must be numbered; and when possible they should be numbered according to the rule once adopted in the British one-inch map: *The figures stand on the contours on the upper side.*

It is somewhat strange that this excellent rule has been abandoned in the later maps of the Ordnance Survey, for it affords a ready means of distinguishing between uphill and downhill. To do so it is not necessary that the map should be examined so closely that the figures of the height may be read. So long as one can see the places in which the figures are, one can see at once which way is uphill, for the figures stand on the contours on their upper sides.

When the country is steep the contours come close together, and it is difficult for the eye to follow them, even on close examination; while if one tries to obtain a general view of the map, the contours merge into one another, and become a mere shade. If this happens in more than a few isolated places it is necessary to guide the eye by accentuating every tenth or every fifth contour. If, for example, the map is contoured at ten metres interval, every tenth contour, that is to say, the hundred-metre contours, should be drawn heavier. This enables the eye to follow the run of the contours, and is a great help in reading the form of the ground. It tends to give the map a stepped appearance, and isolated hills seem to be ridged like oyster shells, as may be seen especially in the 1/62,500 maps of the United States. But this is a small matter in comparison with the real advantage of being able to follow the contours readily.

The worst way of accentuating each tenth contour is to chain-dot it, as in the Swiss maps on the scale of 1/50,000. These dotted contours, which are the only ones figured, disappear except under close attention, and the whole system becomes nearly unreadable. It is instructive to take one of these maps and to ink in the dotted contours, thus rendering them heavier than the rest. Immediately the map becomes readable.

Contours on modern maps are usually printed in colour, brown or red. This is desirable when the map is not layer coloured. But when it is, the contours are better in very fine dotted black lines.

Note that the coast line, or line of high-water mark, is not the zero contour, and should not be coloured in the regular series, which is reckoned from mean sea level. For this reason the coast line is better coloured blue. The true zero contour is never shown.

It is arguable whether contours should be shown on glaciers, which are continually changing. That is a formal objection against contouring glaciers, but the same might be argued against showing them at all. Contours, even if not permanent, have the advantage of distinguishing by their shape between the glacier and the steeper ice or snow slopes which feed it. In the Swiss maps the glacier contours are blue; they are black on bare rock or moraine, and brown on land covered with soil.

We have said that the contour interval must be equal throughout, and that any departure from this rule is very inconvenient. There are certain cases in which it may be legitimate, indeed necessary, to interpolate intermediate contours. For example, in flat country nearly at sea level, such as the fen country in the east of England, there may be a large area which is all comprised within a single contour interval, and in such cases an elevation of a few feet may be more important than a rise of several hundred feet in more elevated regions. Intermediate contours at close intervals are then very useful; but it is evident that they should be readily distinguishable from the contours of the normal series. They may be chain-dotted, or distinguished in some other way. But they should not be drawn exactly like the others.

A serious defect of the older British maps, now remedied, was that above 1000 feet the contour interval became 250 feet, and higher up the interval became wider still. This was destructive of all facility in reading the map. At 1000 feet all the slopes suddenly seemed to become less steep, and it was very hard to train oneself to ignore this misleading appearance. If the objection is made that, in very steep country, contours at uniform intervals come so close together that they can leave no room for any other detail, one may reply that in country so steep as this there is little detail to show, except the contours; and that the closeness of the contours gives the effect of hachuring or hill-shading, with far greater precision than the latter is capable of.

It is desirable to have some guide to the choice of the contour interval. An examination of the best examples of British and foreign maps shows that the rule

Contour interval=50 *feet divided by the number of inches to the mile*

represents pretty well the result of experience. It should be understood, of course, that this rule is entirely empirical, derived simply from a number of maps treated as experiments.

Layer colouring, or hypsometric tints.

The layer system, as it is generally called in English, is a comparatively new system which improvements in colour printing have rendered possible. Its aim is to give to the map the general effect of relief which the contours alone cannot give, because they can be read over only a small piece of the map at one time. A scale of gently graded colour tints is chosen, and all the ground which lies between two certain contours is coloured a certain tint, while that included between the next pair of contours is coloured to the next tint on the scale.

The layer system is exceedingly effective in country which is suited to its use; and it is very often exceedingly unsuccessful in general use. We will examine it in some detail.

It is successful in country which does not require more than seven or eight contour intervals. The colour scale can then be formed of eight tints of one colour, and progress from light to dark or from dark to light will denote progressive increase in height, while even the heaviest tint will not seriously obscure the underlying detail of the map. Where more than eight tints are necessary it is difficult to ensure that the heavier tints shall be transparent enough to leave the detail legible; and it becomes necessary to pass from one colour to another. At this point difficulties begin.

It is not possible to work up in tints of one colour to the point of change, from light to dark, and begin the new colour with the lightest tint, for the contrast is then too violent; the scale must be inverted at the change of colour, and if the first series ran from light to dark, the next must run from dark to light. This inversion of the colour scale introduces difficulties in reading the map, which it seems to be impossible to avoid. They can, however, be diminished by care in choosing the colours of which the scale is composed. It is asserted, and apparently rightly, that the change from one colour to another is least disagreeable when the colours run in their spectrum order. A colour scale which runs from green through yellow to orange and red is more agreeable than one which passes from green to orange without the intervening yellow. How far this principle rests on physiological foundations is not at all clear; but it seems to be true in effect. Undoubtedly the most successful layer-system maps are those which follow this rule in the selection of the colours for the scale of tints.

A more serious difficulty is that even when this system of choosing colours is pushed as far as possible, it is still impossible to provide enough tints to serve for a great range of layers at a uniform contour interval. And were it possible to select the tints, the number of printings required would be prohibitive. It is necessary, therefore, in the higher ground, to widen the interval which is represented by one tint; and this produces the same inconveniences as the wider spacing of the contour interval produces in contoured maps without the layers. This difficulty is felt especially when the country to be represented is a high plateau, such, for example, as the Transvaal. In such a case differences of a hundred feet are as important as they are at sea level; yet if a uniform scheme were applied throughout they would be shown by differences of colour in the latter case and not in the former. It is not easy to see how to overcome this difficulty. A good deal may be done, however, by drawing in the contours at uniform vertical intervals.

All the patterns of layer colouring in common use proceed from green at sea level towards brown and red shades for the higher altitudes. But one may argue that it is misleading to work up in this way from colder tints at low levels to warmer tints at high, suddenly changing to white or pale blue at the snow line. It would seem more representative to proceed from brown into grey and so to white. But this involves a reversal of the scale of gradation at some point and so far no really successful results have been obtained on this method. The problem may be commended to the experimenter.

The difficulty of deciding where to place the snow line is sometimes evaded by leaving all ground above a selected contour uncoloured; but this is untrue to nature, since the line of permanent snow is usually at a different height on the two sides of a range; in the Himalaya, for example, it is lower on the south than on the north, because the supply of moisture comes from the south. And the glaciers, of course, come much lower down, so that the layer colours must be interrupted for them. There is something to be said for restricting layer colouring to the lower ground, and relying upon contours, hill-shading, and cliff-drawing for the representation of the heights.

Cliff-drawing and rock-drawing.

Where the ground is so steep and broken that contours become impracticable, recourse must be had to rock-drawing, which is a difficult art in which few excel. It has been brought to a high pitch of perfection in Switzerland, notably by the late M. Charles Jacot-Guillarmod, whose drawing of the map of Mount Everest made for the Mount Everest Committee, and published with the third of the Mount Everest volumes, is perhaps the most beautiful example ever seen. But excellent results may be seen on many of the Swiss sheets.

Elementary mountain-drawing.

None of the regular methods are suitable for representing mountains in elementary maps, nor is the "hairy caterpillar" which about a century ago displaced the pleasant little drawings of hills in profile which descend from the earliest engraved maps. Logically they are indefensible, since the map should represent the ground in plan, while these hill-drawings are in elevation, and obscure, or should obscure, detail lying on the far side of them. One cannot show a river passing in a gorge between two mountains. Nevertheless they are the most elegant and simple solution of the problem, how to show in a general way the position of mountain masses, and it is worth while to study the use made of the symbol by the early map-engravers, from the Rome Ptolemy downwards.

Combined systems of showing relief.

We have discussed separately the merits of spot heights, hachures and hill-shading, contours, and colour layers. In practice several of these systems are often used on one map.

Contours and hill-shading make an effective combination, and if the contours are drawn strongly they make it possible to use oblique shading without much danger of the slope in the shade looking steeper than the slope in the light.

Hill-shading combined with layers demands great skill in its application, or the shade changes the tint of the layers, and gives a misleading effect. On the other hand, a layer map without hill-shading is very liable to look flat in the higher regions, where the layer interval is large, unless the contours are drawn fairly strongly at uniform vertical intervals, and show through the layers. It is a common mistake in a layered map to draw only the contours which bound the layers.

One often forgets that there is no real need to carry a particular method of representation through the whole range of heights. Thus one may restrict the layers to the lower ground, and rely on contours alone, or combined with hill-shading for the upper and usually steeper ground. If the contours on snow or ice are drawn in blue, the absence of higher layers gets over the snow-line difficulty. But the merits of any combination can be discussed with profit only with the map before us; and when we are tied to a uniform system for all the sheets of a map covering a great diversity of relief, we cannot expect that all the sheets shall be successful.

Maps in relief.

Relief models of the ground have a certain limited use in teaching or in discussing a situation with those who can never learn to read a layer-coloured map. But they are so costly and cumbrous that every effort should be made to dispense with them, except perhaps for exhibitions. With rare exceptions models have been made with vertical scale exaggerated, often absurdly. It is difficult to say that this exaggeration should never be practised, because slopes seen in elevation in the field do look so much steeper than the same slope correctly modelled and viewed from above, so that a true model does not give the same effect as the ground when one is on it. But the exaggeration should certainly be small, and probably the smaller the scale the less should be the exaggeration.

Town signs.

The importance of a town may be denoted either by the size of the sign which denotes its place, or of the lettering which gives its name. And since there are various kinds of importance it is wasteful to use both methods to show the same thing. Generally the administrative importance should be shown by the sign, and the number of inhabitants by the size of the name. But to this rule there are many exceptions. If the scale of the map is large enough to show the actual plan of the town, there are no town signs properly so called, and then the administrative status must be shown by the lettering. Conversely, on smaller scale maps of new countries, an important place, the seat of administration, may have only a handful of people in it, yet it is essential to have its name conspicuous. Thus the rules for England and for Central Africa must be different. In practice it is usually necessary to put the name up a grade in size if the town is an administrative centre or even only of historical importance, and to put it down a grade if it is a residential annexe to a neighbour. Strict application of such a rule as that all towns of over 100,000 people are shown in capital letters of the largest size in the characteristic sheet has produced deplorable results in some of the 1/1,000,000 sheets of Europe.

Sheet margins.

It is important that all margins of sheets should be divided in latitude and longitude, and that the origin of the system of longitudes should be stated clearly.

It is also desirable that the margins should be divided into sections of some convenient unit of length, and that each section should bear a letter or a number, to provide a ready means of referring to a particular region of the map.

It is further desirable that the margins of contiguous sheets should overlap to some extent, so that it should never be necessary to refer to two sheets for the detail of a small district. The inconvenience of having a town or village on the extreme edge of the sheet is sufficiently obvious; yet, until recently, little attempt has been made to provide overlaps in the sheets of any regular topographical series.

On the French 1/50,000 map there is a partial attempt to mitigate the inconvenience of having important detail cut in two by the sheet margin. East and west, and to a very slight extent north and south, there is room to show important detail beyond the strict limits of the sheet; but the principle upon which the irregular boundary is drawn is not clear; and there does not seem to be any advantage over the plan of making all the sheets overlap by a definite amount. The overlap which is sometimes found in the sheets of the Ordnance Survey maps of England is usually due to adjustment of sheet edges to avoid waste of space on empty sea. But in the new and completely redrawn one-inch map, begun in Scotland, there is a systematic overlap which is very convenient.

When sheets are designed to be joined together it is important to arrange that the names on the border should join up properly, yet be complete on each sheet. This is done by writing the balance of a name small in the margin, so that it is cut off when the sheets are joined. It is often convenient, also, to write in hairline characters in the margin the names of countries or counties of which only a small part appears on the sheet, and to label roads and railways which run off the edge with their origin or destination. Some convention is needed, such as that roads which come in from the left are labelled "From...", and those that go off to the right "To...". It is also well to show by an arrow the direction of flow of any stream that crosses the margin.

When sheets are being drawn in series to be afterwards fitted together, much care must be taken that the margins correspond, not only in the precise position of detail, but in the strength with which it is represented. Marginal tracings must be made of each sheet as it is completed and sent for reproduction, so that the next sheet shall fit.

It is highly convenient that the sheet shall be divided up by lines drawn right across it; but whether these shall be meridians and parallels, or a rectangular grid, depends upon questions of projection and choice of sheet lines which must now be considered.

The projection of the map.

The projection of a map is the system of constructing the network of meridians and parallels, and is usually not a projection in the geometrical sense of the term. The network of the globe cannot be transferred to the plane surface of the map without distortion of some kind; but of the dozen or two projections in common use we may choose the one most suitable for the purpose of the map. It may represent distances correctly along the meridians and one or two selected parallels, or distances from the centre of the map; it may sacrifice distance to the true representation of area; or it may preserve uniformity of scale in all directions at any one point (though not the same scale at different points), in which case small areas are shown of their true shape, and the meridians cut the parallels always at right angles.

The variety of projections is evident in atlases of maps in single sheets covering large areas of the Earth's surface. It escapes notice more readily in the large-scale sheets of a topographical series, where it is usually impossible to determine from a single sheet what the projection is. On one such sheet the representation of distance and shape and area is practically perfect, and the small errors of the projection are masked by stretch of the paper. But the choice of the projection has nevertheless an important influence on all the technical work of the computing and drawing offices, and also on the suitability of the map for the use of artillery in warfare of position, with its highly complex organisation of bearing-pickets, sound-ranging, and flash-spotting. It influences also the relation of the edges or sheet lines of the map to the system of meridians and parallels, marked in the margins if not carried across the sheets.

Sheet lines.

The sheet lines of a series of topographical maps may be strictly rectangular, or they may be somewhat curved, at least top and bottom. If the former, the projection of meridians and parallels will become skew to the sheet lines as one gets away from the middle sheets of the series: Thus in the eastern and western sheets of the Ordnance Survey of England the meridians are inclined to the sides of the sheet some 4°. It is evident that the maps are based on a single projection, cut up into rectangular sheets. In the maps of the United States, on the other hand, or the sheets of the International Map, the sides are practically or quite straight, but the top and bottom edges are curves. One may fit the four adjacent sheets to the four edges of any sheet, and the four corner sheets to make a nine-sheet block will very nearly but not quite fit: the angles of the corner sheets are a little too small for a perfect fit. But it is impossible to make a block of more than three sheets by three without the want of fit becoming grave.

These sheets of the second kind are all plotted independently, and are bounded by meridians and parallels. However far one goes east and west the meridians and parallels are always symmetrical to the sheet, and one has gained this advantage at the cost of a certain awkwardness of shape: the sheet lines converge towards the pole, and the sheets get progressively smaller. In the first system all the sheets are rectangular and may be made of a uniform size, and will fit together perfectly in any number; but the projection gets very badly skewed when the map has great extent. For example, in the old 40-verst map of Asiatic Russia the meridians on the eastern sheets were at about 45° to the margins.

Grids.

The rectangular sheets of the O.S. Popular Edition are divided into two-inch squares, which are numbered along the top and bottom and lettered up the sides, so that a simple reference gives the square in which any detail is placed. But although this is a referenced system of squares, it is not properly a "grid", which should provide a continuous system of reference over the whole territory to be mapped, independent of particular sheet lines or of the scale of the map. In a genuine grid system a point on the ground should have one constant and unique map reference, though for convenience of brevity the first figures of the complete reference may be habitually omitted.

From the way in which grids appeared upon maps during the war it is common to think of them as something overprinted upon a map already made; but this is the wrong way about. When the projection of a map is chosen, it is not usually possible to construct it geometrically; one calculates the rectangular co-ordinates of the intersections of meridians and parallels, plots these points, and draws curves through the series of plotted points. Suppose for example that the map is to be on the scale of 1/100,000, or one centimetre to the kilometre. Plot the projection upon squared millimetre paper. The centimetre lines are then the lines of a kilometric grid. As origin of numbering one will take a point away down beyond the south-west corner, so chosen that the centre of the projection will fall at a convenient round number, say 1000; 1000. The projection may then be cut up into any convenient rectangular sheets, and the numbering of the grid lines marked on the edges.

If one had chosen another projection, the calculated co-ordinates of the intersections of the meridians and parallels would have been rather different, and a point defined on the Earth by its latitude and longitude would fall at a different point of the grid, and have a different map reference. The sheet lines for a rectangular map of fixed size would cut the projections in different latitudes and longitudes.

From this two important facts follow, that are rather apt to be overlooked in thinking of grids. First, the area on the ground corresponding to a square on the grid is not necessarily a square on the ground, though if the projection is orthomorphic it will be exceedingly nearly a square; and secondly, the area on the ground corresponding to a square on the grid is not the same area on different projections, whence it follows that the grid reference of a point on the ground is not a quantity belonging to the point, depending on its latitude and longitude, but will differ with the projection employed. Hence it is misleading to number the lines of a grid or the edges of a rectangular sheet in miles or kilometres from the chosen origin of co-ordinates in the south-west, for it is not correct to say that the point whose grid reference is 397·0; 246·0 on a kilometric grid is 397 km. east and 246 km. north of the point on the ground corresponding to the origin of numbering of the grid. The true statement is that on the projection employed the co-ordinates of the point are 397 and 246 centimetres, if the scale is 1/100,000. The centimetre square on the map corresponds very closely to the kilometre square on the ground, especially if the projection is orthomorphic; but 397 cm. on the map does not correspond anything like so closely to 397 km. in a straight line, that is to say to a great circle on the earth. Let us remember always that the grid belongs to the projection, not to the ground.

In the stress of war it may be necessary to overprint one set of maps with the grid belonging to another. We take off from the latter the latitudes and longitudes of the grid intersections and plot these latitudes and longitudes upon the first, joining them up with smooth curves. These curves will not be precisely straight lines, nor will they cut precisely at right angles, though perhaps near enough to both for immediate needs. And even if the grid thus overprinted is not noticeably distorted, it is sure to lie askew on the map whose sheet lines belong to another system, and there will be continual questions whether there is not something wrong with it. It never looks like a sound job until the whole first map is properly redrawn upon the projection of the second.

The rectangular sheet lines of a series are themselves a grid of a kind, and the same arguments that forbade such a series of great extent in longitude are equally against gridding. One may be content if "grid north" differs from true north by a few degrees; but to have the grid inclined 20° or 30° to the meridian is intolerable. On the other hand, to have to start a new grid at every 10° or so of longitude destroys half the value of the system of gridding.

If a continuous system of reference is nevertheless required we naturally fall back upon latitudes and longitudes, and make it easy to read these on the map by subdivision of the projection into much smaller compartments than usual. Thus a map on the scale of one inch to the mile might be divided into minutes of latitude and longitude to form a "mesh", though it would probably be better to abandon minutes and seconds and work in decimals of a degree, or even of the quadrant. The obvious disadvantages of the mesh are that the size of the unit varies with the latitude, and that its angles are not necessarily right angles.

If the grid is to be used for any refined system of map reference it is necessary to have a convenient way of measuring the position of a point, inside a grid square, to the tenth or the hundredth of the unit. This is conveniently done by a printed card with scales at right angles. A complete map reference would then be of the form 126·43; 287·92, about whose interpretation there can be no doubt. On active service it is usual to make a rule that references are always given to a tenth of a unit and that the hundreds of the units are dropped, so that the above would read 264879. Such rules are apt to be changed from time to time, and it is unnecessary to deal further with them here.

An obvious application of the grid system is to large cities such as London. Theoretically it is desirable that the grid should be the national grid; practically it is rather convenient that it should be a special grid with a well-marked centre, such as the cross of St Paul's or Charing Cross, bearing a good round number. All public buildings, theatres, clubs, police stations and post offices, fire alarms, telephone call boxes and what-not could then be conveniently indexed, and private addresses might include the map reference for the convenience of callers.

In a land where the metric system is used, and maps are published on natural scales, it is obvious that the kilometre must be the unit of the grid. But in England, where the units of length have no convenient relations, and where distances are measured indiscriminately in miles, furlongs, chains, yards, feet, etc., there is no obviously convenient unit. In a discussion at the Royal Geographical Society in 1924 on "The Choice of a Grid for British Maps" the three principal speakers advanced equally strong arguments in favour of 10,000 feet, 10,000 yards, and the mile. The grid actually chosen for the new one-inch map is of 5000 yards.

The grid in national defence.

A sound system of grids is as important a part of the preparation for national defence as a good map, and it is important to remember that in some ways the grid system may supersede the map. A brief reference to the history of grids in the war of 1914–18 will illustrate this matter. When the Western Front settled down to trench warfare, the original small-scale maps proved altogether inadequate. As a temporary remedy the 1/80,000 was enlarged and redrawn to 1/40,000 and 1/20,000, but the accuracy of detail was quite insufficient, and re-survey was undertaken, on the framework of the national triangulations: the old French, the incomplete new French, and the Belgian. These were calculated on different figures of the Earth, and the reconciliation of three different systems of coordinates afforded many problems unsuitable for solution in the stress of war. New secondary and tertiary triangulation was made with 5-inch micrometer theodolites, and extended by intersections as far as possible into enemy territory. Principal detail, such as cross roads, corners of woods, etc. was fixed by plane table, and the lesser detail interpolated, first from the manuscript cadastral plans, and later from air-photographs. But the map-detail was never really good enough for the needs of the artillery, especially in location of enemy batteries and targets, and the most interesting survey development of the war was the construction of a system of coordinates depending directly on the revised triangulation, to which all bearing-pickets, observation-posts, and sound-ranging bases were referred. All calculations for position, range, and direction were made on this system, and were independent of the detail shown on the map.

For the convenience of the artillery, accustomed to work in terms of yards, the British maps of Belgium and France had been overprinted with a square grid, in squares of a thousand yards a side. Consequently the squares did not fit the metric dimensions of the sheets, and there was no continuity between one sheet and the next. Hence a map reference had always to begin with the name of the sheet, followed by the number of the square, the letter a, b, c, or d of the quarter square, and finally the rectangular co-ordinates in the quarter square.

The French had had for years a much superior system; their 1/80,000 maps had always borne the co-ordinates of the corners in metres, referred to the Panthéon as origin, on Bonne's projection. This projection is suitable enough for maps not too far east or west of the central meridian, but the distortion of angle corresponding with the increasing obliquity of the meridians to the parallels as one departs from the central meridian, introduced unnecessary troubles into all calculations of direction for the artillery. Therefore the French early in 1917 abandoned the Bonne projection, and introduced a new system based upon a very close approximation to Lambert's conical orthomorphic projection. The unification of the Allied command in the spring of 1918 made it imperative that all maps should carry the same grid. The British maps were being brought into conformity, when the arrival of the armistice rendered unnecessary the completion of the change.

The grid of kilometre squares was numbered from a zero south and west of Paris, so that all numbers were positive. Each map of the regular series on the scale 1/20,000 covered 15 by 10 km., and each grid-line was numbered from the above-mentioned zero, the numbers increasing northwards and eastwards. The last figure of the grid number of the western and southern sides of each kilometre square was printed in the south-west corner of the square to provide for the identification of pieces cut from a sheet. Co-ordinates on this abbreviated system repeated themselves every ten kilometres. To avoid this ambiguity each block of 10 km. square bore a capital letter printed at the corner of the block; and if further definition was required the complete co-ordinates were given, or the number or name of the sheet.

To use this system with advantage it is necessary to start with a well-constructed set of tables giving the rectangular co-ordinates of the intersections of meridians and parallels at small intervals from the origin, so that by simple interpolation any position in latitude and longitude derived from the triangulation can be reduced immediately to the system, and plotted on the sheet. To these points all plane-table detail is adjusted, and from these points all positions of bearing-pickets, sound-ranging bases, and observation-posts for flash-spotting are deduced by resection. All bearings and all magnetic deviations are referred to "grid north", the use of true north being abandoned. The position and zero-line of a battery being determined from bearing-pickets, and the position of the target by sound-ranging or flash-spotting, the bearing and range of any target can be calculated without any reference to map detail, and batteries can be laid on a target without any preliminary registration.

Scale.

The question of scale is of prime importance; the aim of the map-maker is to show as much as the scale on which he works will permit, and he must learn in the first place what are the possibilities of the different scales, and secondly how to make the most of them.

The scale of the map is defined by a statement of the relation between a distance measured on the map and the corresponding distance on the ground. The distance on the map will be measured in one of the smaller units of length, that on the ground by one of the larger; thus, we say that the map is on the scale of one inch to the mile, or one centimetre to the kilometre, or again, we may invert the statement and say that the scale is one mile to the inch, or one kilometre to the centimetre.

There is no very definite rule in English practice, whether to say, one inch to the mile, or one mile to the inch. In general one avoids the use of fractions in the statement: thus, one says one inch to the mile, but four miles to the inch, not one-quarter of an inch to the mile. On the other hand it is common to speak of the equivalent in inches of one mile as the characteristic of the scale, and to speak of the inch map, the half- or quarter-inch map, and so on.

This is the common way of speaking, but it is being superseded by the more accurate method of giving the "representative fraction", that is to say, the ratio of the distance on the map to the distance on the ground. There are 63,360 inches in a mile. Therefore the representative fraction of the one-inch-to-the-mile map is 1/63,360; and the representative fraction of a map on the scale of one centimetre to the kilometre is 1/100,000.

We shall note at once that the British system of measures of length produces awkward fractions: for the half-inch map the R.F. (usual abbreviation for representative fraction) is 1/126,720; and for the quarter-inch map it is 1/253,440. This arises of course from the fact that the number of inches in the mile is not a round number. In countries where the metric system is used this difficulty does not arise, since all the units of length, large and small, are related decimally. It follows that the R.F. for a map of such a country is naturally a round number, and it has become the practice to speak of "natural scales", meaning thereby, scales for which the R.F. is a round number, 1/250,000, 1/100,000, 1/80,000, and so on. The term does not seem to be a good one, since there is not anything natural in the use of a decimal system of units, but rather the contrary, if it is proper to argue from the fact that the metric system, a late creation, has not yet displaced altogether the complicated non-decimal systems which natural man had evolved. Natural scales, in fact, are natural only in countries which have become habituated to the decimal system of measures; in other countries they are not natural, but their scientific convenience has led to their gradual introduction.

We may notice that there is some diversity of opinion as to the choice of a convenient range of so-called natural scales: Shall the series run

1/200,000, 1/100,000, 1/50,000,

or shall it run 1/250,000, 1/125,000, 1/62,500?

There is an undoubted convenience in subdivision by two, so that a sheet on one scale is subdivided into four sheets on the scale next larger, and so on. This appears in either series. But the second series is derived by such steps from the one-in-a-million scale, while the first is not. On the other hand the numbers in the first series are simpler, more round, than the numbers in the second. The divergence illustrates very well the difficulty that continually occurs in the application of any purely decimal system, that for many purposes continual division by two is more convenient than division into so many tenths.

Again, there is the advantage in the second series of scales that it differs very little from the series in inches to the mile.

Half an inch to the mile is 1/126,720. This differs from 1/125,000 by little more than one per cent., or about the uncertainty in the scale of the printed map which is due to the expansion of the paper with damp. Hence in all ordinary use of the map there is no practical difference between the two, and this fact certainly tends to the choice of the series derived from the one-in-a-million by continual division into halves. Moreover, the recent standardisation of the one-in-a-million map, and the convenience of making all other sheets subdivisions of this, must tend to the adoption of the second series rather than the first.

It seems probable, therefore, that the series

1/250,000, 1/125,000, 1/62,500

will gradually be adopted more and more widely.

It is not unimportant to remark that the R.F. is better printed 1/126,720 or 1 : 126,720, than in the fractional form $\frac{1}{126720}$, which requires figures so small that they are read with difficulty.

There can be no hard and fast rule as to the scales suitable for maps intended for definite purposes. But in general one may say that cadastral maps are larger than 1/15,000; topographical maps between 1/20,000 and 1/1,000,000, the larger scales being for military purposes tactical maps and the smaller strategical; while anything on a smaller scale than one in a million can hardly be considered topographical, but may be called for convenience an atlas map.

Construction of scales for maps.

The word scale is used in two senses. In an official manual we used to find on consecutive pages these two statements:

"The scale of a map is the relation between a measured distance on the map and the corresponding distance on the ground."

"The scale of a map should be, for convenience, about six inches long."

In the second sense the scale of a map is the diagram which allows one to translate distances on the map into distances on the ground expressed in any unit desired—not necessarily that employed in the first instance when the map was made. Thus it may be convenient to put on an inch-to-the-mile map a scale of hundreds of yards; or to construct a scale of miles for a French map on 1/80,000.

A scale should be constructed so that any length taken from the map with a pair of dividers can be read off at once from the scale.

Suppose that the scale is to give thousands and hundreds of yards.

The main scale is divided simply into thousands of yards, set off to the right of the zero. To the left of the zero one additional space of a thousand yards is subdivided into hundreds, and if it is numbered at all, which is usually not really necessary, it is numbered to the left, that is to say, in the opposite direction to the numbering of the main scale.

A few moments' trial will show the advantages of this plan. Suppose that the distance is between 3000 and 4000 yards. One leg of the dividers is set on the division 3000; the other leg reaches beyond the zero, and the odd hundreds are read off from the subsidiary scale. On this system the actual figures required are read directly from the scale. On any other system the figures shown by the scale must be modified in some way or other.

Example: For the one-inch-to-the-mile map, construct a scale of thousands and hundreds of yards.

1000 yards = 1000/1760 inches = 0·568 inch.

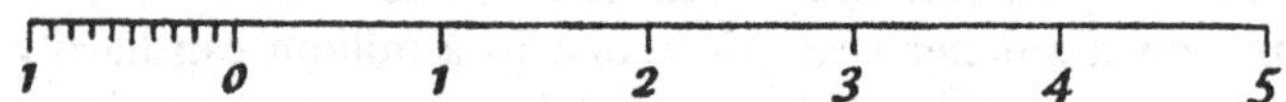

Fig. 1. *Scale of thousands of yards, for the one-inch-to-the-mile map.*

Note that:

(1) If the zero were placed at the extreme left of the scale,

or (2) If the subdivided portion were numbered from left to right,

or (3) If the scale had been constructed for the unit 500 yards instead of 1000 yards,

it would not have been so convenient as in the above form.

The scales given on the Field Service Protractor will generally serve for the construction of any desired scale without calculation.

It is not always sufficient to assume that a single scale of miles or kilometres will serve for the whole map: for maps covering large areas it cannot. See, for example, the scale engraved on the margin of the new Asia 1/4M series (conical orthomorphic projection with two standard parallels).

On this map, the projection being orthomorphic, or the same in any direction round about a point, but varying somewhat rapidly with latitude, the distance to be measured should be broken up into lengths, and each piece measured in terms of the scale appropriate to the mean latitude of the piece. But when the scale varies with the longitude as well as with the latitude, this method is no longer possible. Moreover, in general the straight line joining two points on a map is not necessarily the route of the shortest way from one to the other. Hence in maps of large areas, such as atlas maps, there can be no such thing as an accurate scale of distances.

Problems in the construction of scales.

Exercises in the construction of scales are often set in military examinations, and many rules for the solution of these problems are given in the military textbooks. It seems to the author that these rules, like the old-fashioned multiplicity of rules in arithmetic, defeat their purpose by suggesting that there is a rule to be remembered, instead of a common-sense sum to be worked.

Thus, for example: Given a map on the scale 1/100,000: To construct a scale of miles.

On the scale one mile to the inch the R.F. is 1/63,360. For the smaller scale 1/100,000 one mile is obviously 63,360/100,000 or 0·63 inch, and the scale is easily constructed with a rule graduated to inches and tenths, or with the diagonal scale of the protractor.

Map reading.

Facility in reading from a map all the information that is implied in its conventional representation of the ground is an accomplishment only to be attained by constant practice in the field, and it is useless to attempt to lay down many rules for it. Much excellent advice on the methods of learning the art is given in the *Manual of Map Reading and Field Sketching*; and we will not attempt to cover again the ground which is thoroughly gone over in that book. But we will try to supplement what is said there by a few additional notes.

Sections and profiles.

In the solution of many practical problems it is required to draw a section across the country, from the information which is given on the topographical map. Suppose a line drawn across the map along the line of the proposed section: it will cut the contours at a number of points, and these points will furnish the principal material for the construction of the section. Lay a strip of paper along the line, and mark on it the points where the line cuts each contour. Draw perpendiculars from the edge proportional in length to the heights of the contours above sea level: this will be facilitated if the paper is furnished with equally spaced lines parallel to the edge, and distant from one another by the contour interval on the adopted vertical scale of the section.

A broken line joining the summits of these perpendiculars is a rough approximation to the section, which can be much improved by studying the other indications of height upon the map. Thus, it is improbable that a spot height will fall exactly on the line of the section, but a good deal may be learned from those that fall near. Hill-shading is useful in suggesting the probable shape of the country between the contours, and care should be taken to utilise the information which is given by streams. By taking advantage of these subsidiary sources of information it is possible to construct a fairly good section from a topographical map; and of course the closer the contour interval the more accurate the result.

It is however inevitable that a section drawn in this way should be conjectural in its details, and if accuracy is required it is necessary to run a line of levels across the country along the line of the section. For this process we refer to the section on Levelling (page 202).

A profile of a road is a section run along the course of the road, instead of straight across country. It is constructed in the same way as a section, by laying a strip of paper along successive lengths of the road and marking off the points where the contours are crossed. In addition there are generally a number of spot heights marked along the road, and these are especially useful because they give summits. When a road rises above a contour and falls below it again there will usually be found an intervening spot height, which will mark the summit.

It is often useful to have the profile of a road along which it is necessary to send traffic, and profiles of the main roads of the country are being published for the use of motorists. Unfortunately the makers of these books have sometimes adopted the wrong name "contour road book", instead of "profile road book". This bad mistake should be avoided.

Problems of intervisibility.

Questions on the use of maps very often demand the solution of the problem, whether one point is visible from another. It is evident that this can be solved, in some degree, by drawing a section from one point to the other, and seeing if any intervening point on the section rises above the ray joining the given points. Since the section can be only approximate it is clear that caution must be used when the ray clears the intervening ground by only a little.

In most cases it is not necessary to draw more than a part, if any, of the section, for it is easy to see where are the critical points.

It should never be forgotten that in undulating wooded country the trees are the principal obstacles to mutual visibility of points, and it is not possible to estimate from the map the effect of the trees. Hedgerow trees, which are not marked on the map as woods, since they offer little obstacle to the passage of men, are often a complete obstruction to the sight. Consequently problems of mutual visibility of points are more useful in testing the understanding of the map than in deciding whether it is actually possible to see from one point to the other.

Reconstruction of the view from a given point.

It is an excellent exercise in map reading to try to reconstruct from the map the view of the country from a point of vantage, and to draw it as a panorama. From the chosen point of view radiating sections are run across the country, and it is usually not hard to discover what heights form the sky line in different directions, and what features are prominent in the middle distance. When these are determined, a study of the contours gives a general idea of the shape which these features will present, and then by the exercise of some ingenuity it is possible to construct a panoramic sketch which reproduces the broad features of the view. A few exercises of this kind are very useful in teaching facility in reading relief, especially in town schools where it is difficult to reach actual ground on which to practise.

Identification of distant objects.

It will often happen that a wide view, say in the Lake District, or on Dartmoor, presents a confused panorama of peaks which cannot be identified by their relations to one another as shown on the map. To identify one such distant object it is necessary to take its bearing by compass, subtract the westerly deviation, as given on the margin of the Ordnance Survey map, and with the protractor to lay off a line on the resulting true bearing from the point of observation. For the instrumental process reference may be made to page 145. The distant object will then lie on this line drawn upon the map; and of the several objects at varying distances which may chance to lie on it, it will not in general be difficult to select the right one.

The true bearing is laid off from the meridian through the place of observation, and it is important to remember that this meridian is not necessarily parallel to the sides of the sheet, which are not "practically north and south", as they are sometimes said to be, but may even in a small country like England be several degrees out of the meridian. Neither is the grid which may be on the map necessarily north and south. But a map should always, and does usually have its margin divided in latitude and longitude. The meridians formed by joining the points of equal longitude will usually converge, but any intermediate meridian may be interpolated with sufficient accuracy for our present purpose, and on topographical sheets any meridian may be considered a straight line.

The reverse problem sometimes arises. It is desired to identify on the ground, or at least to discover in what direction it lies, an object which is marked on the map, but which is not apparent to the sight. In such a case draw on the map a line from the point of view to the object in question, measure with the protractor the true bearing of this line, and add the compass deviation west. Then with the compass to the eye turn slowly round until some object is found which has the given compass bearing. The object sought will lie in the same straight line from the observer, and with this indication it is often possible to discover it, or at any rate to find exactly where it lies.

Measurement of areas.

The first consideration which arises in this problem is: Are the areas correctly represented on the map; or is the projection on which the map is constructed an equal area projection?

Topographical maps are not usually constructed on a projection which is theoretically an equal area projection; but the misrepresentation of areas will always be much less than the uncertainty which is introduced by the expansion and contraction of the paper with damp.

Atlas maps will very often be on projections which misrepresent areas grossly; but for a consideration of this question reference must be made to a treatise on the somewhat intricate subject of Map Projections.

An easy way of measuring areas on a map is by superposing a sheet of tracing paper regularly ruled in small squares. First count the number of whole squares embraced by the area, and then estimate how many whole squares more are equivalent to the array of partially included squares along the boundary of the area. From the scale of the map calculate the area represented by one square of the tracing paper, and thence derive the whole area of the figure in question.

More elaborate methods involve the use of special instruments such as the planimeter.

The "Amsler" planimeter is a beautiful instrument whose use is simple, but whose theory defies explanation in an elementary book. It consists essentially of an arm carrying a pointer which is moved round the boundary of the area to be measured, and a wheel which rolls on the surface of the paper. A second arm is pivoted to the first between the pointer and the wheel, and its other end is free to turn about a fixed point. As the pointer moves round the boundary the wheel revolves, sometimes forwards, and sometimes backwards. The nett amount of rotation forwards is measured by a revolution counter and by divisions on the drum attached to the wheel. The nett rotation, multiplied by a suitable constant, gives the area.

Style in lettering.

The map-engravers of the best period, round about 1600, used two characters, Roman and Italian, treated with great freedom in tails and flourishes wherever it was desirable to fill a space more completely than the unadorned letter would do, but preserving a unity of style. The letters they engraved were not all precisely to pattern, being subtly varied to suit the occasion just as the spacing has always to be varied a little to avoid other detail, which is of course the reason why typed names with their mechanical regularity never look right on a map. With the gradual decay of beauty in printing to its lowest point about the middle of the nineteenth century, the style of engraving on maps became continually more fixed, and approached most closely to the printer's type just when that was at its worst. Its characteristics are an exaggerated difference between the thick and the thin strokes, and exaggerated serifs—the little cross strokes or tails at the beginnings and ends of the principal strokes which are all important in giving it character and legibility. With the exaggeration of the serifs came the base practice of ruling them in and joining up letters by their serifs so that they became confused one with another: of which the worst examples may be found in some varieties of the ugly sloping character known as "stump".

About the same time came in a taste for variety, of which we may distinguish two classes: variety apparently for its own sake, as in the characters employed on the older Ordnance Survey maps, with ordinary roman, capitals and lower case; italic capitals; thickened hair-line capitals and lower case, without difference in thickness of the strokes, but with large serifs; the so-called egyptian or block letters without any serifs at all—the plainest (though far from the most legible) sometimes curiously called "grotesque"; and stump; or variety with intent, that towns, rivers, mountains and capes, railway stations, political divisions, old regional names, tribal names, ruins, and what not, shall all be distinguished one from another by the style in which their names are written. In this system, which is elaborated to the highest pitch in the characteristic sheet of the one-in-a-million International Map, relative importance is distinguished by size, and the nature of the object named by the character of the lettering.

The student should examine carefully the characteristic sheet of the International Map (Paris Conference, 1913) to learn how very much information can be conveyed by careful variation in the lettering of names: most of which escapes the uninstructed user of the map. To make full use of the possibilities requires, however, a higher standard of draughtsmanship than is easily acquired: and when the necessary skill is reached, there remains the more exacting requirement of artistic judgment in selecting the appropriate size of lettering, of spacing out, and of placing so that the map is a harmonious composition. A common defect is to spread the spaced-out names too widely: they should clear the boundaries of the area they occupy by a space decidedly larger than the space between the letters. Another common defect is to interlace spaced-out names so that the map seems covered with letters scattered at random.

A great part of this deliberate variety is superfluous, since there can rarely be any doubt about the character of the feature to which the name is attached, if the name is suitably placed. The conventional sign should tell the tale, not the character of the lettering. One can scarcely doubt that the beauty of the older engraved maps depends more on the beauty of the form of the lettering, and the restriction to two naturally related styles, than on the mechanical perfection of the engraving, which was not in fact of the machine-made exactness that the modern engraver or draughtsman achieves.

While opinions may differ on the question of beauty, there can be no disputing the fact that present styles of map lettering are so expensive that few books are published with adequate maps. This is the strongest argument for a radical change in style, and the change must come by adopting a letter that can be drawn in single strokes of the pen, without elaborate building up, except for the largest names. There is here much room for study and experiment.

A device which has been often used, but never successfully, is to stamp the names on the drawing with type, or to print the names from printer's type on slips of paper and stick them on to the drawing. The experiment has perhaps never been properly tried, because until lately there were no types available of suitable form, and especially there were none that would stand photographic reduction without losing the correct proportion between thick and thin strokes. But typed names can never have the necessary flexibility to fit themselves to the curves of rivers, or to avoid detail, which the draughtsman achieves by almost imperceptible adjustments of spacing and position. If it were possible to make a machine that would stamp singly in exactly the place required any one of a carefully graded number of founts of type; this want of flexibility might be overcome; and the matter is perhaps worth the attention of inventors. But the easier solution is to abandon the present style of lettering, except possibly for the most expensive style of map, and to substitute what is nearer fine writing than printing.

In the opinion of the writer, these are the lines on which we already see an improvement in the style of map drawing. But many will not agree, and it is true that though the above principles have to some extent been proved sound in simple drawings for sketch maps, especially for blocks in the text, they have never yet been tried on elaborate topographical maps in many colours. The opposite school of thought and practice is well established, and its methods will not be soon altered.

Some first attempts at a new style were made in the line maps in the text of the *Geographical Journal* in the latter half of 1926, with an italic letter based on the engraved character of the Hondius map of 1608, recently reproduced in facsimile by the Royal Geographical Society from the unique copy in its collection, and since that date all the blocks in the text and some of the lithographed maps have been produced in this style.

The new Ordnance Survey alphabets.

About the same time Captain Withycombe and Mr Ellis Martin of the Ordnance Survey began to experiment with the design of a new alphabet for the projected new edition of the one-inch map of which the first sheet was published on 15 August 1931 and is analysed on page 64. On 12 November 1928 Captain Withycombe read a paper at the R.G.S. on "Lettering on Maps" (*Geog. Jour.* LXXIII, 428, May 1929) which was followed by an interesting and amusing discussion. In this paper Captain Withycombe traced the development of our alphabets from the Roman capitals of the classical monuments, especially the Trajan Column, through the Roman text or lower case derived from the former in the monastic texts by writing more quickly with a pen, to the Italic alphabet based on the pointed Italian handwriting. For three hundred years maps were engraved on copper plates, and about the end of the eighteenth century the capitals became narrower in proportion to their height, fine hair-lines and thickened downstrokes lost all relation to one another, and the serifs became so exaggerated that instead of emphasising they concealed the distinctive forms of the letters. In the lower case also hair-lines too thin to support the downstrokes were introduced, and the fine balance of the old letters was lost. Just at the time then when lithography displaced engraving on copper the style of lettering was least suited to the new process; and the later adoption of photographic processes made matters worse. Fine lines became broken in reproduction and narrow spaces filled up. It was essential to design new alphabets which while suitable for photographic reproduction had also intrinsic decorative qualities and harmony of effect.

The capitals of the new alphabets are based on Trajan, Leonardo, Dürer, and the mediaeval type-designers. The narrow S, the wide M, O, and W, and a hundred more subtle points all make for

A A B C D E
F G H I J K L M
M M N O P Q R
R S T U V W W
W X Y Z & &
abcdefghijklmn
opqrstuvwxyz
1234567890

A B C D E F G
H I J K L M N O
P Q R S T U V
W X Y Z & &
abcdefghijklmn
opqrstuvwxyz
1234567890

Alphabets designed for the new One-inch Map, Ordnance Survey.
Drawn with steel pen, built up.

A B C D E F G
H I J K L M N
O P Q R S T U
V W X Y Z
a b c d e f g h i j k
l m n o p q r s t u
v w x y z
1 2 3 4 5 6 7 8 9 0

A B C D E F G
H I J K L M N
O P Q R S T U
V W X Y Z
a b c d e f g h i j k
l m n o p q r s t u
v w x y z
1 2 3 4 5 6 7 8 9 0

Alphabets designed by R.G.S. Redrawn by C. E. Denny 1944.
Single strokes of quill, except for serifs.

Plate VIII

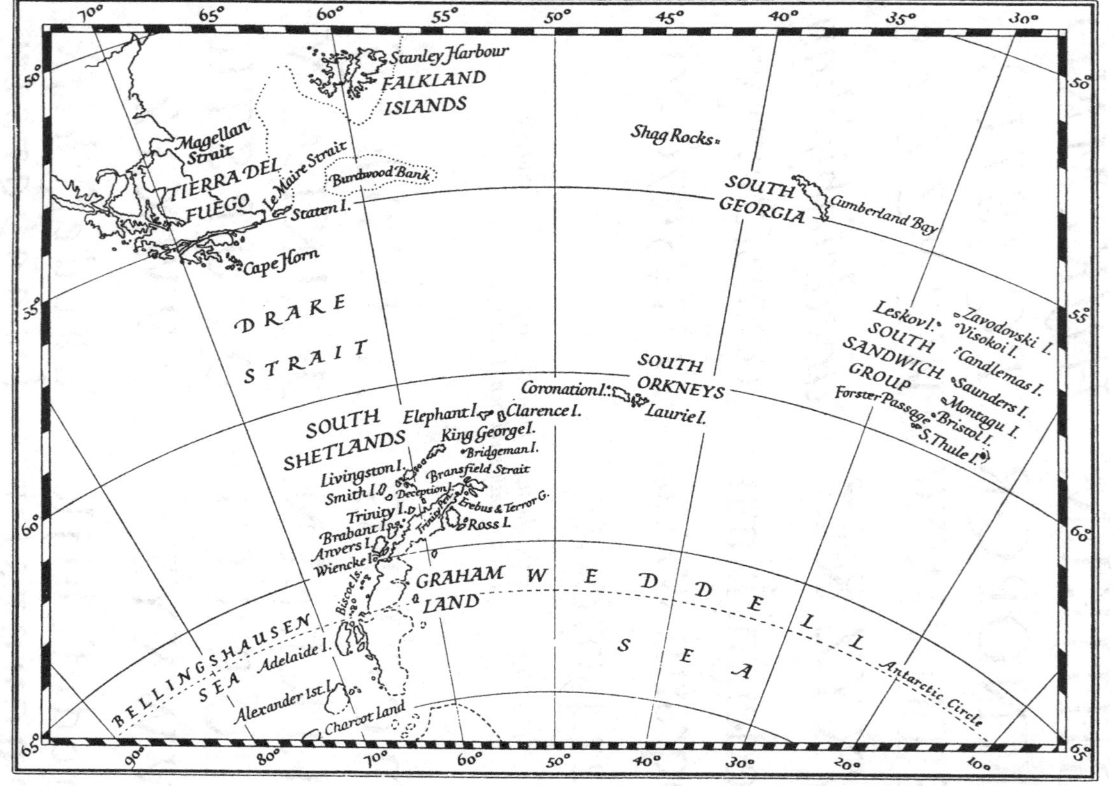

(*Geogr. Jour.* LXXIX, 178, *March* 1932)

Falkland Islands and Dependencies. Quill-written Italic lettering.

legibility and beauty. In particular the white spaces within the letters are well designed. The lower case alphabet is based on pen strokes, and grace is obtained by the natural tapering of strokes to a point instead of by a violent change from thick downstroke to hair-line. In the Italic text the distinctive form of the *g* and other letters has been restored, and the letters are finished with a strong tapering stroke which reproduces well. But the problems of a National Survey are peculiar to themselves and freedom of style suitable for single sheets or maps illustrating books and journals would be out of place on a series of official maps, in which the work of many draughtsmen has to produce a uniform result.

The preceding paragraphs are condensed from Captain Withycombe's exposition of the designs by Mr Ellis Martin for the new Ordnance Survey alphabets, to be drawn carefully with fine steel pens on hard white Bristol board, for "helio" reproduction. All except the smallest letters perhaps are built up, the thick strokes being outlined by two fine strokes and then filled in. The speed of drawing is of the order of twenty names per day, which is too expensive for anything but maps of permanent importance.

The first object of the R.G.S. experiments was to find a style more economical of time, which is done by avoiding built-up letters except in the larger and more important names. All the work is done with fine quills. Where letters must be built up, the quill is easier and quicker, and the result not so mechanically perfect, but perhaps more beautiful. All but the largest letters are written freely in single strokes, with the quill held in one position throughout. The speed is about five times that of the steel-pen drawing. An amateur draughtsman cannot hope to make anything like the finely finished steel-pen drawn letters; but with a little practice with quills he can quickly learn to write the quill-pen letters. Incidentally he will much improve his ordinary handwriting at the same time.

In the discussion on Captain Withycombe's paper there were strong differences of opinion. There were those who thought that the twentieth-century Englishman knows more than the early Roman; that present-day lettering is a survival of the fittest; that the worst man to choose to design letters for maps is an artist; all that is wanted is legibility and simplicity. To which of course the reformers replied that the new alphabets submitted were as simple and legible as could be made; that the subtleties of design are introduced to get legibility; that a letter written in a single stroke is at any rate simpler besides much quicker to make; and that, if not for the Ordnance Survey at least for everyone else, the ruling consideration is economy. Maps drawn in the old style are too expensive for books and journals.

The effect of a map depends greatly on the frame. The division of the border into the units of latitude and longitude may be made decorative by blacking in alternate divisions, as in Pl. V and VIII, and careful attention to the style of the numbering. All the reference, explanations, scales, etc., should be included within the frame, and the title be placed either in the top of the frame or in a panel on the face of the map. Students should be advised to study the best examples of seventeenth-century engraving, but must at all costs avoid deliberate archaism, which has a deplorable effect.

The spelling of place names on maps.

The subject presents great difficulties owing to the very different values given to the letters of the Roman alphabet in different European languages, and the inadequacy of this alphabet for the representation of the finer shades of sound; to the use of arbitrary combinations of letters whose value cannot be derived from the values of the individual letters, such as the *cz* and *sz* in Polish; and to the use of accents and diacritical marks on the consonants in Czech (properly Česky), Serbo-Croatian, and Romanian. It is hard to get agreement on the method of transliterating other alphabets into Roman, especially when as in Arabic they are rich in consonants which sound almost alike to European ears, and cannot be distinguished in Roman without the addition of diacritical marks. There are naturally great diversities in the transcription of native names, as African, into European languages. There are peculiar difficulties in dealing with Chinese names, in which each character (ideograph) represents a syllable pronounced very differently in different parts of China. There are conventional spellings contrary to accepted modern rules, but so well established that it would be pedantic to alter them, as Warsaw, conventional for Warszawa which is pronounced Varshava; and there are other conventional names which are badly corrupt, or versions of names not now used in the place, as Damascus—a pseudo-classical version of Dimishk esh Sham, now called simply Esh Sham—or Constantinople for Istanbul, itself a Turkish corruption of Greek. Finally, there are many cases in which not the spelling so much as the whole form of the name is in question, as Monastir, the conventional Turkish for Serbo-Croat Bitolj, or Kluj in Transylvania called Kolozsvar in Magyar and Klausenburg by the Austrians. The last class of difficulty usually comes from change of ownership, and has been much aggravated by recent events in Europe.

In descriptive writing it would be pedantic to substitute the true for the conventional names of very well-known places, but a map should usually give the true name according to the practice of the responsible authorities, with conventional names following in brackets. Thus we may continue to write of Bozen, but should spell it Bolzano on our maps. It is clearly disadvantageous for the traveller to find that the names on his map do not correspond with those written up at the railway stations.

The following principles have been adopted for the International Map: The spelling of any name in an independent country or self-governing dominion shall be that adopted by the country or dominion. The spelling of any name in a colony, protectorate, or possession shall be that adopted by the country governing it, if it uses the Roman alphabet. The spelling of names in China shall be that adopted in the Chinese Postal Guide.

These rules are well adapted to the International Map, but require modification for general maps published by any one country. For example, the common Arabic word Wadi would be spelled Ouadi or Oued on a French map of Algeria, but in a British map of that French colony would be spelled Wadi or Wed. And generally, native names transliterated or recorded by French or German or Portuguese must be transformed by regular rules (such as those developed by Knox in a well-known War Office pamphlet) into the British system of transliteration before they can be used on British maps. But the process is not very certain, since one language may not be really capable of expressing the sounds of another: for example, when a *w* is found in the German version of an African name, it may be intended for the sound of the English *v* or *w*, and endless mistakes are caused by this defect.

The nationality of the map-maker must affect also the treatment of descriptive names, especially those containing topographical terms like river, mount, cape. The English have a good way of compromising. We say neither Cap Grisnez nor Cape Gray Nose, but Cape Grisnez; on the other hand, we say neither Mount Blanc nor White Mountain, but Mont Blanc. Others are more thoroughgoing in translation. The French say Cap de Bonne Espérance, and the Russians call the Grampians the Grampianski Gori. These simple examples show the difficulties of making any rules.

The R.G.S. system.

The phonetic system used very widely in English-speaking countries for the transliteration of non-Roman alphabets and in the reduction to writing of unwritten native names is that devised more than half a century ago by Admiral John Washington, long known as the R.G.S. system, and lately revised by the Permanent Committee on Geographical Names for British Official use, as the R.G.S. II system. Its leading principles are that the vowels are pronounced as in Italian and consonants as in English, that every letter is sounded, and that no superfluous letters are retained. These principles are not however quite sufficient to cover all cases. For example, there is no vowel in Italian corresponding to the sound of the French *eu* or the German modified *ö*, and in general there is trouble with diphthongs. It is also necessary to give arbitrary values to certain combinations of letters, as *dh* to represent the sound of *th* in "this"; *kh* to represent the hard guttural in "loch"; to use *q* for the second *k* sound in Arabic; and so on.

The Permanent Committee on Geographical Names for British Official use.

The foreign place names found in British books, atlases, and newspapers have in the past been spelled chaotically, partly because they have been written with English values for the vowels, partly because they have been taken from foreign books without the necessary transformation. The difficulty became so acute during the war that on its conclusion the Admiralty approached the Royal Geographical Society with a request that it would form a Committee including representatives of the principal Departments of State, to publish authoritative lists of place names for British official use. The Committee has now been at work at the House of the Society for thirteen years, and has already published many lists, while a very extensive series of tables of the alphabets of the world transcribed into the system R.G.S. II has been published for the Committee in the *Technical Series* R.G.S. With the official spelling of each name there is given where necessary the pronunciation and stress, the name in the original character if not Roman, the form used in foreign countries, and (in brackets) the forms which are definitely wrong. The lists are therefore a mine of interesting material.

The Committee, however, has not been able to secure consistency. It has had to accept as conventional many false spellings so well established that to alter them would be impossible: such names as Florence, Leghorn, Flushing, Dunkirk. A more serious inconsistency arises from the existence of several established but different systems of spelling Arabic names, in India proper, in the adjacent countries, in Mesopotamia (now Iraq), and in Egypt. The Committee could not undertake the impossible task of converting those who had taken the trouble to make a definite system and stick to it, but has been content with the less heroic work of establishing as much order as possible out of the existing confusion. It accepts without question the decisions of the *Imperial Gazetteer* for British India, of the Geographic Board of Canada for all place names within that Dominion, of the Survey of Egypt for places within its zone of interest, and so on.

A distinct branch of the Committee's work is to decide between alternative names for a place. There are, for example, many islands in the Pacific which bear several names, perhaps one European and two or three native names. The latter confusion is often due to mistaking a district name for that of the whole island; and very often the island as a whole has no native name. Mountains are often known by different names to the people living on different sides of them; but when these names are descriptive, they often express the same thing in different languages.

Misuse of personal names.

Explorers during the last century, as well as earlier, often gave personal European names to mountains and lakes in Africa, and to a less extent elsewhere, without troubling even to discover whether there was an established native name. This is now generally recognised as a mistake, and there is a healthy movement towards expunging these names from the map. Probably a few of the more modern will survive, like Victoria Nyanza and the Victoria Falls; and there is no chance now that contemporary native names, even if they could be discovered, would ever displace the Dutch names in South Africa such as Drakensberg, Vaal, or Orange. Towns new founded, on no ancient settlement, may more legitimately, though not always with advantage, bear European names such as Harrismith, Pretoria, Brazzaville, Elizabethville; but natural features should bear native, or native-sounding names, as Ruwenzori, whose separate peaks and passes should also be native sounding, even if names have to be invented for them, rather than that they should bear the British and Italian names assigned to them by the illustrious mountaineer who first climbed them. Happily the present Governments of East Africa are fully alive to this question, and are reverting to native names, especially in the Tanganyika Territory.

In the Himalaya there is one European name—Mount Everest—and the Survey of India are resolved that there shall never be another, but retain provisional designations as K 2, N 53, when there is no certain native name, or else use the serial number of the peak on the sheet, as Pk. 8/72 I. There is, however, a real difficulty. An exploring and climbing party with a strong interest in a special neighbourhood naturally want names for each individual peak and satellite, which are very important to them, though of no importance to the general nomenclature of the Himalaya, with its ten thousand permanently snow-clad peaks. The Survey of India nomenclature is impossible in descriptive narrative, and one must sympathise with travellers who have found it necessary to give temporary names like Bride Peak, though they cannot be allowed to attach actual personal names, however august. Sir Halford Mackinder set a good example when he called the principal peaks of Kenya by the names of two legendary Masai chiefs, Nelion and Batian. The Mount Everest Committee were faced with a formidable difficulty in preparing the map of the magnificent region so suddenly disclosed. A few genuine Tibetan names were found, such as Chomolungma for the main group, Gyachung Kang, and Cho Uyo. Useful descriptive names like North Col and South Peak were translated into Tibetan as Chang La and Lhotse. The history of one name is instructive. When Mallory first saw the splendid mountain we now call Pumori he wished to call it Mount Clare, after his daughter. This could not be allowed; but Pumori means The Daughter Mountain, and his intention is thus happily fulfilled in a legitimate way. In the Shaksgam region Major Mason has followed the same principle, though generally by getting his Ladakhis to suggest names themselves. And he has at last found the native name for K 2. By a slight mispronunciation of the Survey appellation it is Cheku. It remains for someone to turn into euphonious Balti the uninspired neighbouring Broad Peak and Staircase.

In utterly uninhabited lands, as the Antarctic, there can be no native names, and here personal names are less objectionable, though they have been overdone, especially in calling every little piece of coast a Land.

Pronunciation and stress.

Pronunciation glossaries form an important part of the marginal information on a map, whenever the names are in Roman script with other than the normal values for the consonants. British maps have done well in this respect for Eastern Europe, but until the P.C.G.N. took the matter in hand there have been no attempts to make a guide to the sounds of Welsh, Gaelic, and Erse names, and pronunciation glossaries are still absent from the sheets of Great Britain and Ireland in the International 1/M series—perhaps because English names defy such treatment. In any case, such glossaries are of limited value, for a foreigner may arrive at more or less correct pronunciation of each syllable without getting the stress and intonation that are in some languages so important.

The rules for transliterating names into the Roman script necessarily require a letter-for-letter transliteration, with no freedom to recognise peculiarities of pronunciation either in particular words or particular districts. Analogous rules for transliterating British names into Russian or Arabic would not distinguish the different values of the vowel *a* in Cambridge, Camborne, Swanage, Thames, Bath, Durham; nor the characteristic differences of the vowel in the Hampshire, Lancashire, and Scots dialects. No system of phonetic spelling can give the finer differences; but the broader are within the scope of a relatively simple system such as the R.G.S. II: and we should often arrive at a better representation of the sound if we wrote the name in this system without reference to the way it was spelled in the original. For example, a transliteration of Tibetan names produces terminal consonants that are not actually sounded. But a serious objection to writing phonetically names which have an original spelling is that they are then often unrecognisable by scholars. For example, if Chinese names were spelled phonetically it might be impossible to recognise the similarity of two names actually written in the same characters.

Stress is quite as important as pronunciation, and the lists of the P.C.G.N. are careful to indicate the stress whenever it is unusual or commonly mistaken, as in Himálaya or Saráwak. Allied to stress is the interruption in the pronunciation of Arabic names indicated by the *hamza*.

Irreversibility of transliteration.

It is vain to argue, as some pedantic scholars do, that it should always be possible to recover the original characters by reversing the transliteration. Since the Roman alphabet is much less rich than some others, such as Arabic, it must be reinforced with special signs, marks, or new characters, to provide enough equivalents. These extra signs cannot be printed without difficulty, and are quite unfit for use on maps.

It is much worse with Chinese, which may have twenty or thirty different ideographs that must be represented by one and the same syllable in transcription. Reversibility is thus quite impossible: nor does the transcription give much guide to the pronunciation, owing to the great differences in the spoken language in different parts of China. Moreover, the Postal Guide sometimes deliberately introduces irregularity in transcription, to distinguish two places of nearly the same name. One should always remember that, however badly the transcription may represent the sound, a letter or telegram addressed in the Postal Guide system will reach its destination: otherwise it will not.

Sheet names.

The system of the International Map provides for each sheet a reference letter and number, and a name. It is generally found that the sheet is referred to by its name, not its number: and therefore the name ought to be that of a city near the centre of the map: not necessarily the most important name. But the number is required for reference to subdivisions of the sheets, on larger scales.

For example the numbers of sheets on 1/250,000 must show their relation to the 1/M sheet to which they belong. It would be convenient if the sheet number could contain in some way the latitude and longitude covered by the sheet: but this does not seem to be practicable. The well-known series of the Austrian 1/200,000, covering 1° each way, was centred on the intersections of the meridians and parallels of the degrees, and in practice it is not convenient to remember that the sheet whose latitude is given as +46° covers from 45° 30′ to 46° 30′.

Sheet numbers based on the 1/M map.

The sheets of the 1/M series cover 6° of longitude by 4° of latitude. They are numbered eastward from the meridian opposite that of Greenwich, and lettered north and south from the equator. Thus the sheet covering Central England is North N–30.

Sheets of the 1/250,000 series, measuring 1° each way, are lettered in four rows, A to F, G to L, M to R, S to X, 24 to each 1/M sheet. Thus the sheet of such a series including Cambridge would be numbered North N–31/S.

Sheets of the 1/125,000 series, measuring 30′ each way, are numbered in Roman numbers I to IV. The Cambridge sheet would be North N–31/S–III.

Sheets of the 1/62,500 series, measuring 15′ each way, are lettered N.W., N.E., S.W., S.E.; and the Cambridge sheet would be North N–31/S–III–S.W.

This method is applied to the G.S.G.S. series of Africa. It is systematic: but no one could call it convenient, and few could remember readily the numbers of the sheets adjacent to a given sheet. A principal defect is the mixture of letters, Arabic and Roman numerals, and points of the compass. The method does not provide for a series on the scale 1/500,000, which may well be required some day.

Map reproduction.

We cannot in the scope of this book describe the many processes which are employed in the production of maps: but a slight sketch of the principal methods is essential to accurate judgment of the results.

Wood engraving, used in the first printed maps, very soon gave place to engraving on copper, which has lasted to our day, but is now little used for new work, though the plates survive as the basis from which many maps are printed by transfer to stone or zinc. The engraved copper plate could not give very many impressions without becoming worn: hence the very great difference between successive editions of the older atlases. The process of steel-facing an electrotype of the original enabled the Ordnance Survey in the last century to produce maps in black as perfect as early impressions from the original plate, but they had to be printed slowly on a hand press. And printing in several colours was practically impossible. When, therefore, it was desired to produce a map in several colours, transfers from the black original were taken on stone or zinc; the parts of the map not required in the particular colour were erased, and additional plates were prepared to print tints not represented by anything on the original. Thus until lately the Ordnance Survey maps in colour were largely printed from plates prepared on transfers from the original copper.

It was possible to make considerable alterations on an engraved plate, and electrotypes of the originals could be cut up and pieced together to make new plates of different sizes from the old. But this process has recently reached its limits, and when it became necessary to make a fresh start with the one-inch map, it was decided that the new original should be drawn on card, and the printing plates made by photography.

The needs of the official surveys or of atlas publishers are not the same as those of more temporary publications. In the middle of last century the maps for books or journals were generally on copper, but as this became more prohibitively expensive they were, and still often are, drawn upon stone, from a carefully prepared copy, by a lithographic draughtsman. The maps, for example, of the *Geographical Journal* are still mostly produced in this way, which has certain advantages. It is not difficult to make corrections on the stone; and the copy from which the lithographer works need not be highly finished; it is sufficient if the material is all there in the right place. Moreover, the process can be practised in modest establishments, and is not extremely costly. But it introduces between the original and the published result the personality and the rather mechanical style of the commercial draughtsman, so that it is hard to secure small refinements and changes in style.

Modern practice in the large establishments, such as the Ordnance Survey, tends more and more to photomechanical reproduction from finished drawings on paper as the surest method of combining ease of revision with speed and accuracy. In the old process of lithography the map was drawn in reverse upon the stone in a greasy ink, which repelled the water with which the stone was kept moist, but took the ink from the inking roller. In the modern process of "zincography" a finely grained zinc plate is preferred to the stone, as less expensive and more convenient, though not permitting such large editions. The problem of making the zinc printing plate is to reproduce the drawing in grease upon the zinc, either to scale or to a chosen reduction.

In either case the zinc plate is coated with fish-glue sensitised with ammonium bichromate, which on exposure to light becomes insoluble. If the drawing is to be reproduced on the same scale, it is made on white drawing paper or tracing paper, pressed face downwards tight against the coated plate, and exposed to strong light. The film protected from light by the lines of the drawing remains soluble, and is washed away, leaving bare lines on the zinc. The plate is then inked up with greasy ink, and the remaining film loosened and washed away with dilute acid. This leaves the drawing in ink on the plate, and impressions are taken as in ordinary lithography.

If the drawing is to be reduced in scale a negative is made by the wet-plate collodion process, and the sensitised zinc plate exposed under the negative, rendering the lines of the image insoluble. The plate is then inked with greasy ink and the soluble part of the film dissolved away by warm water and gentle rubbing, which leaves the image in greasy lines on the zinc plate as before, ready for printing.

The former process is called the Vandyke, the second the Helio process of map reproduction.

The above description of the Ordnance Survey methods is condensed from a paper by Lieut.-Colonel Craster, R.E., on "Photomechanical Processes of Map Reproduction" (*Geog. Jour.* LXV, 301, April 1925) to which reference may be made for further details.

The plates which are to print lines in colour are made thus from the original negative, and the lines which are not required on each plate are "duffed out". The plates for printing flat or stippled and ruled tints are made by exposing under screens and stopping out the parts not wanted. A single plate can print about five tints, stipples, single line, crossed line, and solid. And by overprinting one may get eleven or twelve tints from two plates. But really good layer colouring is very hard to do. The specimen plate attached to the report of the International Map Conference of 1913 was printed at the War Office. It is a masterpiece of layer printing, and it is not too much to say that none of the published maps have quite come up to the specimen colour scale or "gamme".

Chapter III

BRITISH OFFICIAL MAPS

The Ordnance Survey.

THE official mapping of the British Isles dates from the rebellion of 1745. When military posts were established in the Highlands after the defeat of the Young Pretender at Culloden, the Deputy Quarter-Master General, Lieut.-General Watson, made an elaborate compass sketch with hachured hills, which was afterwards extended over the whole of Scotland, and was the first official map of our country. The original sheets are in the British Museum, on the scale 1/36,000, and a reduction on a single sheet was published by General Roy, who had taken a large part in the work, in his *Military Antiquities*.

In 1765 Roy was appointed Surveyor-General of Coasts and Engineer for making and directing Military Surveys in Great Britain and in 1783 began a triangulation between London and Dover for the geodetic connection of the observatories of Greenwich and Paris, measuring the base on Hounslow Heath in the following year. This was the beginning of really accurate survey in England. On Roy's death in 1790 the Duke of Richmond, Master-General of the Ordnance, resolved that the work thus begun should be continued, and on 10 July 1791 appointed two artillery officers to carry on the Trigonometrical Survey. The Ordnance Survey dates, therefore, from 1791, and its first map, of Kent and southern Essex, in four sheets on the scale of one inch to the mile, was published on 1 January 1801.

The six-inch-to-the-mile survey was begun in Ireland in 1825, and the first published sheet, of Londonderry, is dated 1830; the publication was completed in 1846. The six-inch map of Great Britain was begun in 1840 and the twenty-five-inch in 1854.

Until about thirty years ago the Ordnance Survey produced excellent maps but appeared to have little interest in the question whether they sold or not. The maps were published in the form of flat sheets only, and before they could be used out of doors it was necessary to spend at least as much as the price of the map in having it mounted on linen and put into covers. The means of distribution and sale of the maps were also very inadequate.

About the time when the original pattern of maps printed in black only from engraved plates was superseded by editions printed in several colours from transfers of the engraved plates, maps mounted on linen and folded in covers were placed on sale, and later the method of mounting was introduced by which the map could be opened at any desired section with one hand, while the other hand remained free. The map did not catch the wind nor offer much surface to the rain, and it could be consulted without attracting attention. These advantages were secured by folding the map face outwards instead of inwards, and cutting the cover into two parts, so that there was no hinge at the back as in a book. This simple expedient did more than anything else to increase the convenience of using a map, yet strangely enough it was not imitated, and it was always difficult to get an unofficial map-mounter to understand how to do it. Stranger still, the Ordnance Survey themselves abandoned this excellent practice: it seems to have been a casualty of the war; and when it was recently re-introduced it was considered a novelty.

It is a mistake to buy for field use any maps but those mounted on linen, and folded without dissection. The dissected map is more flexible and lasts longer; but it is impossible to make measurements upon a dissected map, and that should be sufficient to condemn it.

The Ordnance Survey maps.

Ordnance Survey maps are at the present time in a state of transition, and it is a little difficult to make a summary that shall not be out of date in a few years.

The standard map is the one-inch-to-the-mile, which was originally engraved on copper and printed on a hand press from electrotypes of the original plates, in black only. The beauty of these maps has never been surpassed, but the method does not lend itself to printing in colour, and especially not to the printing of flat layers of colour so much employed to show the relief of the ground in modern maps. Hence the various coloured editions of the one-inch map have been prepared by transferring from the engraved original in various ways on to lithographic plates, with separation of the detail that was required to be in different colours. The present

'Popular" one-inch map is the last that will be produced in this way.

It was recently decided that the time had come to prepare a new original, and that the process of engraving on copper should be abandoned for drawing on card, from which the various colour plates on zinc are prepared by photographic methods (see page 59). It is thus easier to make alterations and keep the maps up to date. The black plate is somewhat heavier than the former engraved plate, on which the finer detail was scarcely strong enough to show through the superposed colour. The result, first seen on the new Scottish sheets of the "Popular" one-inch map, was a considerable improvement.

A short while before the war a beautiful new type of one-inch map was produced in a few sheets—e.g. the Snowdon area—which was delicately layer coloured, hachured in purple, and contoured; a few sheets are published as Tourist maps. After exhibiting many combinations of layer colouring, hachuring, and hill-shading, the Ordnance Survey then for a time abandoned hachures and hill-shading altogether, and upon the Popular one-inch map rely on contours only, but at fifty-feet intervals, for representing relief. On the new one-inch map of England, analysed below, we have again the hachures, in two colours, and a simple layer colouring, superimposed on the contours.

The question is often asked: How do British maps compare with those of other countries? The answer is that while in a few specialities some foreign maps may excel those of the Ordnance Survey—the Swiss in rock-drawing; the Norwegian in a very attractive style of hill-shading—there is no other country with so large a range of excellent maps on many different scales, and particularly, there is no other country with anything like the large-scale series on 1/10,560 and 1/2500, and the town plans of five and ten feet to the mile. These large-scale maps, however, hardly concern geographers.

The topographical maps are

The 1/63,360, or one inch to the mile, which is the standard topographical map of the country, and in its recent form on large sheets is the most valuable for local use.

The 1/126,720, or half-inch map, which used to be the standard military map for war and manœuvres, and is the most generally useful for motoring and bicycling.

The 1/253,440, or quarter-inch map, which is the most useful for planning long distance journeys, and for intelligent railway travelling.

The smaller scale maps, ten miles to the inch, and the one-in-a-million map, are for strategical and general purposes.

The student should make a point of analysing examples of the different series, since in no other way is it possible to obtain an idea of the possibilities of the different scales, or of the difficulty of finding a uniform system of mapping that shall be satisfactory in all varieties of country. The following notes on selected sheets are given merely as illustrations of some of the more obvious points of interest.

Ordnance Survey of England and Wales. One inch to the mile, 1/63,360. Sheet 144. Fifth (RELIEF) Edition: Plymouth.

Printed in colour at the Ordnance Survey Office, Southampton, 1931. Size 29 by 17 inches plus margins.

This, the first sheet of an entirely new map, has not disappointed the hopes raised by Captain Withycombe's description and illustration of the new alphabets designed for the use of the Ordnance Survey. The new alphabets, Roman and Italic, give to the map a distinguished appearance; the names are legible, the smaller far more legible than before, and also beautiful. The improvement in legibility of numerals from adoption of the "old style" forms is remarkable.

Relief by contours in brown at vertical interval 50 feet, combined with the hachures of the old engraved map in orange overprinted with the same hachures on south-eastern slopes in grey, equivalent to but more delicate and expressive than combined vertical and oblique hill-shading; and in addition a pale layer colouring in buff changing tint at each 500 feet. Figures inset in the contours in brown. Frequent spot heights in black.

Roads classified much as before, first class filled reddish brown, second class buff, with Ministry of Transport numbers. Parish boundaries have been re-introduced, to confusion with footpaths. Railways with much improved symbol giving them due importance on the map; stations by circle or rectangle filled reddish brown.

Woods outlined, with tree symbols in black, and pale green tint; public parks paler green tint, private parks uncoloured; National Trust areas outline chain-dotted, tinted pale green with darker ribbon-border.

Rivers solid blue, very conspicuous; tidal waters and sea pale tint of same colour; contours in sea-bed much more detailed than before. Coast line black with careful detail. Forts and dockyards which were shown on the Popular Edition now again removed.

Projection Transverse Mercator with central meridian 2° W. Grid of 5000 yard squares (2·84 inches) numbered from origin far west and south: the south-west corner of sheet has co-ordinates 800,000; 1,160,000. Margin divided to minutes and intersections of meridians and parallels at each five minutes marked by black crosses. Inclination of grid to meridian stated on eastern margin, with compass variation.

This Relief edition of the one-inch has been abandoned; no more are to be printed, and copies of existing stocks should be secured for collections. New sheets of the Fifth edition omit the delicate layer-colour and hill-shading from the old hachure plates, which made so distinguished an effect. The sheets published up to 1941 are mostly for south-western and southern England; for the northern parts of England the Fourth (Popular) edition has contours at vertical interval 50 feet in orange and frequent spot heights; rivers in solid blue, conspicuous; railways inconspicuous compared with roads; woods with bright green tint; names in black, roman capitals and lower case, and smaller in stump. The sheets of Scotland, more recent than the English, redrawn on the Cassini projection, have contours in brown instead of orange, thickened at each 250 feet; lakes in solid blue like the sea, not lined; submarine contours in fathoms. The black plate is rather heavier than on the English sheets and holds its own against the colours; it is especially good in the large urban areas.

Special District and Tourist maps on the one-inch scale have various styles of hill-shading; and there are a number of sheets on this scale of the Land Utilisation Survey, made under the auspices of the London School of Economics and the Geographical Association, directed by Dr Dudley Stamp and printed in colour on the one-inch map as base: see *Geog. Jour.* LXXVIII, 40 and LXXXI, 541.

The London Passenger Transport Board map on two inches to the mile is made from the drawings for the blue and black plates of the Fifth edition one-inch map. The new style of Ordnance Survey alphabets can be well studied on this larger scale.

Ordnance Survey of England and Wales. Half-inch to the mile, 1/126,720.

The half-inch series is also in a state of transition. On most sheets we have

Sheets of the regular series 24 by 19 inches of engraved surface.

Rivers in blue, double lines filled when wider with hatching or dots; or fine single lines, rather light to be effective among other detail. Lakes lined blue; sea and tidal rivers in pale blue dots.

All names in black: italic capitals except for cities, two grades of roman lower case, and stump; mostly drawn, but some of the older sheets with stamped names in the smaller sizes.

Relief by brown contours at 50, 100, 200,...—1000, 1250, 1500,.., the break in interval at 1000 feet making the hills look flat-topped. Layer colouring two tints of green to 200, then ascending tints of brown.

Reference by 2-inch squares numbered and lettered, special to each sheet, not part of continuous grid.

There are larger District Maps made up from plates of the regular series, but margins in the new O.S. style, with 5000 yard grid. The Peak District sheet has brown layers above 500 feet, but no green below. The Cotswold's Special District Relief Map, 1931, is on the Transverse Mercator with 10,000 yard grid, in the style of the one-inch Relief Map for border, with brown vertical and purple oblique hill-shading from shaded drawing instead of engraved hachures, and strong blue rivers, but made up from older plates, not fully redrawn. The large sheet of Greater London is drawn in the new O.S. character and a good example of its beauty. These three sheets should be studied together.

The half-inch map of Scotland is similar, but the outlying northern sheets are not layered.

Ordnance Survey of Great Britain. Quarter-inch to the mile, 1/253,440.

The fourth edition, completed about 1935, is built up into sheets of varied size, from the plates of the previous edition, with border of the new pattern and 10,000 yard grid.

Relief by contours drawn and figured in brown at 200 feet interval, and layer-coloured in brown, changing tint at 200, 4, 6, 8 and thence at each 400 feet; spot heights only on summits, and none below 200 feet.

Roads classified and numbered by Ministry of Transport, class A red, B yellow, with numbers in red; other roads uncoloured. Railways unclassified, rather inconspicuous; stations in circle filled red. Woods in green tint. Rivers in double or single line bright blue; gradation into tidal waters well managed. Names in Roman and italic capitals, roman lower case, and stump.

The map is designed especially for motoring; road continuations and junctions, official numbers, destination and distance well shown in the margins; traffic diagrams of principal towns bound up with the folded sheets in covers.

A Civil Air edition with special air information overprinted in red, and the main physical features emphasised, is published in twelve sheets arranged rather differently from the ordinary edition.

Ordnance Survey Aeronautical Map, 1/500,000.

Sheet lines based on International 1/M map, but each sheet extended half a degree south and west to overlap the adjoining sheet. Contours at 400 feet; major altitudes (summits and aerodromes) in red; some minor in black. Layers in violet at contour intervals. Place names in black in the new O.S. character; aerodrome names in red, and all heights, in an ugly sans-serif, the numerals peculiarly bad. Town areas in black with roads through them in brown, and railways through them in black edged with white. Principal woods green.

Special air information includes aerodromes and landing grounds, air beacons, radio-stations, and lines of equal magnetic declination. No grid; margin very lightly divided in latitude and longitude, obscurely figured.

An excellent map, a little spoiled by minor defects in design.

Ordnance Survey: Ten mile map of Great Britain, 1/633,600. 1936.

Transverse Mercator grid, 50,000 yards.

No contours, but relief by layer colouring in green, buff, and orange changing at 200, 4, 6, 8, 12, and so by 400 feet. A few spot heights. Submarine contours at 5 and 10 fathoms.

Margins in the new style, but body of map in the older, apparently transferred from the older edition. The margin is numbered and lettered as reference to the squares, discouraging to the proper use of the grid for reference.

Ordnance Survey: Ten miles to the inch. Civil Air Edition. 1933.

No contours, but relief by vague layer colouring in light brown.

Main drawing the same as above, but sheet lines a little different; the graticule in purple hence not quite symmetrical.

Air information overprinted in blue.

Ordnance Survey: Ten miles to the inch. Road Map of Great Britain. 1932.

Simplified and redrawn on the Transverse Mercator projection with 50,000 yard grid: sheet lines not the same as above.

Buff ground tint and relief by brown hill-shading. No heights.

Names well drawn in the new O.S. character.

Roads in red with Ministry of Transport numbers in red and violet, and distances in miles between towns or marked points.

The two sheets may be had mounted back to back in the Ansell folding.

This Ansell fold may be used to advantage on any single sheet map which is too tall for convenient handling. Divide it by horizontal lines into eight equal parts lettered A to H from top to bottom. Mount the sections C, D, E, F on one side of a sheet of cloth, and mount the others upside down on the back of the sheet in the order H, G, B, A from top to bottom. Then if sections

D, E, F are folded backwards behind C the sections A and B appear above C; and if the sections C, D, E are folded backwards behind F the sections G and H appear below F. A map much broader than tall may be dealt with in a similar way in vertical sections; but those on the back are not then mounted upside down.

Ordnance Survey of Great Britain: 1/Million, 15·78 miles to the inch.

The old engraved map of the United Kingdom published in 1904; England, Scotland, and Wales revised in 1926. Two sheets 24½ by 35 inches.

Relief by brown hill-shading; no contours or heights. Rivers, roads, and railways all in black. Coastal waters tinted blue.

Minimum error conical projection with rectified meridians. Meridians and parallels at 1° interval carried across the sheet.

Ordnance Survey of Great Britain: 1/Million.

Great Britain in two sheets, in the style of the International Series, of which six sheets are published by the Ordnance Survey, but Kent falls on the Paris sheet.

On this same scale of 1/Million is the Physical map of England and Wales, and a Magnetic edition of the same; the Population map of Great Britain in two sheets based on the 1931 census, layer-coloured to show the density of population; and three Period maps: Roman Britain, Britain in the Dark Ages, and Seventeenth-century England, besides the Edinburgh and Aberdeen sheets of the International map of the Roman Empire.

Proposed changes in Ordnance Survey Maps.

The Report of the Departmental Committee which was appointed in 1935 has been described as a new charter for the Ordnance Survey. Its principal recommendations were summarised by the Director General of the Survey under three headings. What he calls its basic recommendations are five: (1) that no alterations should be made in the scales of existing maps, or in the styles of small-scale maps (though it approved a trial of a map of 1/10,000 [probably mistake for 1/12,500] in place of the 6-inch map 1/10,560); (2) that all maps, large- and small-scale, should be on national instead of county lines, that is to say, that they should be on one projection for the whole country; (3) that a grid should be shown on all Ordnance Survey maps; (4) that the 1/2500 should be revised continuously; and (5) that a new map on the 1/25,000 scale should be introduced.

Consequent upon these recommendations it must follow that the large-scale survey will be completely overhauled to eliminate present discrepancies at the boundaries of projection systems and errors introduced in the course of its previous revisions, and that the staff of the Ordnance Survey shall be greatly expanded and the work of revision adequately maintained. Owing to the first World War and subsequent efforts of economy it had fallen very much into arrear.

The most striking recommendation of the Committee was that the metre shall be the unit of measurement for the grid, from which it follows that each 1/2500 plan (25 inches to the mile) shall cover 1 kilometre square and each 6-inch plan 5 kilometres square. The large-scale maps are to be square, built up from the kilometre-squares of the basic map 1/2500. It does not necessarily follow that the small-scale maps will be or need be square, but they will probably correspond to an integral number of 6-inch sheets.

These and other, subsidiary, recommendations were discussed at a meeting of the R.G.S. on 16 January 1939 (*Geog. Jour.* XCIII, 314–32), in which there is a full and interesting account of the reasons that governed each recommendation and an almost complete concurrence in the recommendations by representatives of many different interests who took part in the discussion.

Ordnance Survey Large-scale Maps or Plans: 1/2500 and six inches to the mile, 1/10,560.

These large-scale maps or plans do not fall into the category of topographical maps, but the former is the basis of all smaller scales, and has always been on the "natural scale" of 1/2500, which is only roughly 25 inches to the mile. At the above-mentioned discussion the Director General spoke of the advantages of scales in binary or decimal relationships, 1/2500, 1/12,500, 1/25,000, 1/50,000, 1/250,000. There was some reluctance to give up the six-inch, 1/10,560, but the Committee recommended a trial of 1/12,500, which might in time replace it, and definitely recommended a new map on 1/25,000. It did not recommend any change in the one-inch or smaller scales, though 1/62,500 is so nearly one inch to the mile, and its binary divisions so close to the half-inch and quarter-inch that there would be great convenience in use if the change were one day made: the grid squares, for instance, would be one or the other of two standard sizes, and each a round number of millimetres.

The Projection and Grid of Ordnance Survey Maps.

A Map Projection is defined in a formula by which the meridians and parallels and any series of points defined by their latitudes and longitudes on the Earth's curved surface may be transformed into a corresponding set of lines (straight or curved) and of points upon a plane. The rectangular coordinates of the intersections of meridians and parallels, and of any other points, are calculated from the formula of the projection and plotted with respect to the coordinate axes for which the formula is constructed.

For the Fifth edition of the One-inch map the Ordnance Survey adopted the projection called the Transverse Mercator, and all maps as they are revised are being calculated on that projection. More recently it has been decided that all coordinates shall be calculated and plotted in metres. This transverse Mercator is calculated for a central meridian 2° west of Greenwich, and from an arbitrary origin on that meridian in latitude 49° north. The rectangular coordinate axes of the projection are therefore a straight line corresponding to the meridian 2° west, and a line at right angles to it through the point of it corresponding to 49° north. With these axes, all points west of the central meridian would have one coordinate negative. To avoid this change of sign, and to secure that the other coordinate for the whole mainland of Great Britain shall lie within a range of 1000 km., which is just conveniently possible, the quantity 400 km. will be added to the east-west coordinates, and 100 km. subtracted from the north-south coordinates. Thus all the coordinates may be described as Eastings and Northings from the transferred origin, though in fact the north-south axis through this new origin does not run true north, since it is parallel to the central meridian, not convergent to it.

We now want a convenient system of reference for points on the map, and find it better to have a rectangular system than references in latitude and longitude. So we draw two sets of parallels to the coordinate axes, at the same interval, naturally in a round number of metres, forming a series of equal squares: the grid. And since the original axes of coordinates lie far outside most of the sheets, and only a part of the coordinate system falls upon any single sheet, one does in fact construct for any sheet a grid which is an element of that system, and plots upon it the lines of the projection corresponding to the meridians and parallels, or at any rate their intersections with the bordering lines of the grid; one plots the trig. points and other points whose coordinates have been calculated, and fits the detail into these ruling points. We can then say for short that the map is drawn upon the grid, but not that the grid is drawn upon the map: at least in the regular procedure above described.

But during a period of changing over from one projection or one central meridian to another, it will often be necessary to show upon the old maps the grid lines that properly belong to the new. These can easily be plotted, and serve perfectly well for map references. But the transferred grid lines will be a little curved, and the compartments not truly square, though the defect may be inappreciable. In such a case, something very like a true grid is drawn upon the map.

To transform the yards grid into the new kilometre grid.

Many recent Ordnance Survey maps bear a grid in yards from a transferred origin a million yards to the west and south of the true origin, and until the change over to the metric grid is complete it will often be necessary to interpolate the new metric grid on a map gridded in yards, the new grid being numbered from the transferred or false origin 400 km. to the west and 100 km. to the north of the true origin. A table for this purpose, based upon figures communicated by the Director General of the Ordnance Survey was published in *Geog. Jour.* XCVIII, 118. (For table see p. 285.)

Maps of the Geographical Section, General Staff.

The Geographical Section of the General Staff, War Office, publishes a very large and varied selection of maps of British territories, and of other parts of the world in which Great Britain has special interests, or for which it is not likely that maps will be made by the local governments. Some of these maps—e.g. Cape Colony and the Orange River Colony series—represent work of the Colonial Survey Section; others are compilations from route traverses and miscellaneous work, and are provisional in character; others again, as the early sheets of the Canadian map, were published by arrangement with the Dominion authorities. As the organisation of the British dominions beyond the seas progresses, they naturally tend to take over series begun by the G.S.G.S., and set the latter free to devote attention to important general series, such as the International Map on 1/Million, the 1/Two Million of Africa, and the 1/Four Million of Asia.

The style of production of the maps of the G.S.G.S. is varied, and some of the earlier provisional issues were comparatively rough. When the building of the new War Office gave them adequate accommodation their printing office speedily gained a high reputation for layer colouring, and some of their productions—notably the specimen scale of layers for the 1/M map attached to the report of the Paris Congress of 1913—have never been surpassed. The specimen scale, indeed, set a standard too high for successful imitation: but students should examine it as an example of what colour printing may be.

A few of the recent maps of the G.S.G.S. are analysed below.

Asia. 1/4,000,000. Geographical Section, General Staff.

Sheets covering 30° of longitude by 20° of latitude, and thus of different sizes, the southern very large. Conical orthomorphic projection with two standard parallels 27° and 63° north lat.

Relief by contours in brown at 200, 500, and thence by 500 metres and layer coloured in green, brown, and red. Water and water names blue, and coasts outlined with blue ribbon. Lakes solid blue, sea blue lined. Roads and railways black. Horizontal scales of miles and kilometres showing the great variation of scale with latitude.

This map was planned by the R.G.S. during the war (and one sheet drawn for reproduction) on the scale 1/5M, not 1/4M, and between latitudes 20° and 80°: hence the dimensions five times those of the 1/M sheets, and the choice of standard parallels. When after the war it was taken over by the G.S.G.S. it was reproduced on 1/4M and extended to the equator, while no sheets of the top row 60° to 80° have yet appeared. This explains why the projection and size of sheet seem to be so ill-chosen, and the names written too large. But it is a very valuable map now nearly complete, and the earlier sheets under revision.

Two supplementary sheets of the East Indies on Mercator's Projection.

(Sheet 22, Mongolia: 1/M sheets K to O, 46 to 50. 1921.)

Africa. 1/2,000,000. Geographical Section, General Staff.

Sheets covering 12° by 8°, on alternate sheet lines of the International Map, and the same projection.

Brown contours at 200, 500, and thence by 500 metres. Most sheets layer coloured on the International colour scale adapted. Water blue. Rather overburdened with large district and tribal names.

(Sheet North C 36, 37; B 36, 37: Abyssinia. 1925.)

The compilation of this series was begun at the R.G.S. during the war and the earlier sheets published by the G.S.G.S. without layers. By arrangement with France, Egypt, and the Sudan the whole of Africa will be covered on a uniform plan. The British sheets nearly complete; the others in progress. Note the useful diagrams in margin: indexes to boundaries and to adjacent sheets. The French sheets extended to 10° of latitude.

(Sheet Algérie. Service Géographique de l'Armée. Édition provisoire. About 1922.)

Asia Minor. 1/250,000. Geographical Section, General Staff. No. 2097.

Sheets mostly 21½ by 17½ inches, covering 1½° of longitude by 1° of latitude.

Relief by brown form lines and very few spot heights. Water and water names blue; woods in green tint.

Compiled from various sources and published in thirty sheets between 1908 and 1919. Being superseded by new series with contours at 50 metre intervals, conforming to the new sheets of the Eastern Turkey in Asia series.

(Sheet Angora. 1919.)

Eastern Turkey in Asia. 1/250,000. Geographical Section, General Staff. No. 1522.

Sheets mostly 20 by 17½ inches.

Relief by brown form lines at 250 feet interval, with 1000's strengthened. Rivers blue, woods green tint. Fifty sheets published from 1902 to 1919.

A new series redrawn from the Turkish Staff Map 1/200,000 in hand, with contours at 50 metre interval, every fifth strengthened. G.S.G.S. No. 1522A.

(Sheet 5: Trebizond. 1924.)

Syria. 1/250,000. Geographical Section, General Staff. No. 2321.

Sheets 25 by 17½ inches or irregular.

Relief by brown contours at 50 metre interval, every fifth strengthened. Water blue; woods green tint.

Redrawn from Turkish Staff Map 1/200,000.

(Sheet Latakia. 1924.)

Iraq. 1/250,000. Geographical Section, General Staff. No. 3723.

Sheets about 14 by 17½ inches covering one degree of latitude and longitude.

Relief by contours in brown at 250 feet interval, every fourth strengthened. Water in blue.

(Sheet North J–38/T: Mosul. 1921.)

China. 1/50,000.

Relief by contours at 20 metres, strengthened at the 100's and supplementary 10, 30, 50 dotted. 1000 metre grid.

(G.S.G.S. 3789: Shanhaikuan. 1927. Grid I.)

China. 1/250,000.

Brown contours at 100 metres. Water blue. 10 km. grid No. 2. (G.S.G.S. 3825: Hongkong and Canton. 1927.)

China. 1/1,000,000.

Brown hill-shading. Inland water blue, sea green, roads red. (G.S.G.S. 1936: Shantung, 1905, covering 9° by 5°.)

No hill-shading. Water and sea blue.

(Sheet Chih-li South, 1909, covering 8° by 5°; Ssu Ch'uan East, 1908, covering 8° by 5°.)

The great variety of maps published by the G.S.G.S. makes it impossible to do justice to them in a textbook. They are well described in the Catalogue of Maps published by the G.S.G.S. with full indexes. To the lists of African series are notes such as "Maps on the scale of...are published by the Survey Department. ...These maps, as they appear, supersede the corresponding G.S.G.S. publications".

The Survey of India.

The Survey of India has a long and honourable history, and no country of the world has contributed more to the advancement of geodesy, the organisation of topographical survey, or the methods of work in difficult frontier country. Very great improvements in map reproduction have been made in recent years, though the standard of draughtsmanship and the choice of type for names still leave something to be desired. The Survey of India is very enterprising in extending its surveys, or if that is not possible, its compilations, far beyond the frontiers, and the best of its native surveyors have made for themselves a high reputation in plane-tabling difficult mountain country.

To cover a country so large as British India with maps on topographical scales takes many years, and during the progress of the work the style of reproduction necessarily varies with the development of printing processes, the ambitions of the Surveyor-General, and the requirements of economy. We can do no more than analyse briefly a few sheets typical of the various styles and epochs.

India. 1/63,360. (One inch to the mile.)

Helio-zincographed at the Survey of India Offices, Calcutta. Sheets about 15 × 17 inches, covering 15′ each way.

Relief in brown contours at 50 feet interval, combined with and lost in hachures in hill country. Railways black, roads red, both well classified. Town and village symbols solid red. Perennial water blue, non-perennial water-courses black. Characteristic signs attached. Names in black (water names blue), typed and ugly. Meridians and parallels at 5′ interval across sheet; margins not divided. Longitudes from Greenwich. District boundaries heavily hand-coloured.

(Sheet No. 38 L/15: N.W. Frontier Province. 1909.)

Others similar to above, but not contoured. Water names in black. Town plans in red stipple. Cultivated areas yellow; woods green.

(Sheet No. 53 K/12: United Provinces. 1918.)

Others similar to the first, but with oblique hill-shading added in purplish grey. Water names black. Towns and villages shown by groups of red dots.

(Sheet No. 43 J/4: Kashmir and Jammu. 1912.)

Assam. 1/63,360. Survey of India.

Relief by brown contours at 50 feet interval, approximate, every fifth strengthened, and grey oblique hill-shading from north-west. Names typed and small for scale. Villages in red dotted groups; roads red; perennial water blue, non-perennial black; cultivation yellow; woods black tree symbol and green tint.

(Sheet No. 83 M/4. 1926.)

Earlier sheets of the series without the hill-shading.

India. 1/126,720. (Half-inch to the mile.)

Generally similar to the 1/63,360. Sheets same size on half the scale, covering 30′ × 30′.

Relief by contours at 100 feet interval, incomplete, very faint, and confused with hill drawing. Spot heights rare on plain, frequent in hills. Oblique hill-shading in grey. Perennial water pale blue, non-perennial black. Glaciers in blue lines. Woods green tint, cultivation yellow. Tree symbols black. Town sites red stippled. Boundaries of provinces and districts hand-coloured with heavy ribbon. Meridians and parallels at 5′ intervals carried across sheet. Margins not divided.

(Sheets 44 M/N.E.: Punjab, 1920; or 52 D/S.E.: Kangra District, 1923.)

Iraq and Persia. 1/126,720. Survey of India.

Sheets about 14½ by 17 inches covering half a degree each way, divided into 2½ minute "rectangles".

Relief by brown contours at 100 feet interval, every fifth strengthened, or by brown form lines, according to the state of survey. Roads and tracks red; permanent water blue; cultivation yellow; woods green.

Sheets very much scattered, according to the exigencies of military survey, and rather rough. One of the few maps bearing a latitude and longitude "mesh", since discarded.

(Sheet 2 J/S.W.: Karind, 1920; or 138 N/N.E.: Tikrit, 1925.)

India. 1/253,440. (Quarter-inch to the mile.)

Sheets same dimensions as the 1/63,360 but on one-quarter the scale, and covering each 1° × 1°: hence the common name of Degree Sheets.

Relief by contours at 100 feet interval in brown, every fifth strengthened, and vague hill-shading in brown or grey. Spot heights in large type. Railways black, roads and tracks red. Perennial water blue. Town signs black. Meridians and parallels at each 15′ carried across map. Margins not divided. Names all in black, sans serif, too uniform in size, typed. District boundaries heavily hand-coloured.

Another edition of same date and at same price, layer coloured in well printed shades of green, yellow, and brown. A good map rather disfigured by heavy green boundaries of forest reserves.

(Sheet 84 K: Burma, 1915; or 43 G: Rawalpindi, 1915.)

India and Adjacent Countries. 1/1,000,000. Survey of India.

Sheets covering 4° each way, same dimensions as above.

Relief by brown contours at 250, 500, 1000, and by 500 to 4000, then two steps of 1000, two of 2000, and thence by 2500: a needless diversity. Vague oblique hill-shading in grey from various directions. Towns black, well classified; roads and tracks red; glaciers and perpetual snow in blue; water blue; sea contoured in blue. Names carefully drawn, and style much better than the larger scales.

Earlier sheets had no indication of perpetual snow or glaciers.

Another edition layer coloured in green, yellow, and brown.

Early sheets of this edition left the country above 16,000 feet uncoloured as snow-covered, conventional and quite incorrect.

(Sheet 58: Ootacamund, 1922; 52: Leh, 1925.)

Note that the Survey of India is publishing also sheets on the same scale conforming to the sheet lines and style of the International Map.

India. Southern Asia Series. 1/2,000,000.

Sheets bounded by meridians and parallels; each 2° carried across. Normal dimensions covering 12° of long. by 8° of lat.; some sheets 16° by 8°.

Relief by contours in brown at increasing interval, layer colouring as in the 1/M series, and highly conventional oblique hill-shading in grey.

Railways black. Roads red. Rivers blue. Place names in black, typed; water names in blue.

A valuable provisional series, but not so good as the 1/M.

Another edition of same date without layer colouring, but with political boundaries emphasised by coloured ribbons.

(Sheet Northern Persia. 1914.)

Maps of the Dominion of Canada.

For a long time the surveyors of this Dominion were occupied in cutting up the Dominion Lands into square townships, and true topographical survey was neglected. This is now happily changed. Canada has a Geodetic Survey second to none; its topographical surveys are making steady progress, if somewhat slow in comparison with the size of the area to be covered. It was early in the field with its own particular method of photographic survey in the mountainous country of its boundaries, and in the development of rapid methods of air survey. Its principal need is a layer-coloured map of the whole country on a small scale.

Canada. 1/63,360. Department of Militia and Defence.

Sheet about 24 by 17 inches, covering 30′ by 15′.

Brown contours at 25 feet interval, every fourth strengthened. Spot heights numerous. Roads yellow; water blue; woods green tree symbol. The whole drawn very light and rather difficult to read. A series begun about 1907 by the G.S.G.S. for the Department, but since engraved and printed in Canada.

(Sheet Quebec: 1920. Dept. of Militia and Defence.)

Canada. 1/63,360. Topographical Survey.

Sheets on same system of sheet lines as above, and printed at Ottawa.

Contours in brown very faint, at 100 feet interval, each fifth strengthened. Roads carefully classified in red and yellow; forest green tint. Division of the land into townships six miles square shown in grey, numbered in ranges east and west and townships north and south, all divided into half-mile plots. Full characteristic sheet. Margin ticked into four-mile grid, and number in corner, but difficult to follow as there are other ticks.

(Sheet No. 52–M–V: Kamloops. 1926.)

Canada. 1/126,720. Department of Militia and Defence.

Sheets same size as above covering 60′ by 30′.

Brown contours at 25 feet interval, every fourth strengthened, and some sheets layer coloured. Water blue; woods green symbol.

A few sheets only, produced by the G.S.G.S. before 1914.

Canada. 1/190,080. Topographical Survey.

Sectional sheets covering 15 by 8 townships.

Contours brown at 100 feet to 5000 and 250 above, each fifth or fourth strengthened, printed rather light and figures hard to read. Roads filled red and yellow; water blue; woods green tint.

A beautiful map somewhat spoiled by the sectional division, seemingly inappropriate in the Rocky Mountains. The greater part of the Dominion Lands covered by these sheets either in outline or with full topography as above. The rôle of the principal meridian and the relation to it and to the parallels of the townships and ranges is a curious study.

(Sheet No. 164: Banff. 1925.)

Canada. 1/253,440. Topographical Survey.

Sheet covering 2° of longitude by 1° of latitude.

No contours; occasional spot heights. Water blue; woods green tint; marsh or muskeg grey. Interesting as the first regular topographical series made largely by air survey.

(Sheet No. 47–Q–R: Pointe du Bois. Provisional 1926. The boundary between Ontario and Manitoba divides this sheet. Ontario has a four-mile grid numbered from an unspecified origin, perhaps part of the United States system. Manitoba has the township division of the Dominion Lands. The two systems meet oddly upon this sheet, which is well worthy of study.)

Maps of the Commonwealth of Australia.

A great deal of primary triangulation has been done, but until lately the efforts of its various survey departments were devoted too much to surveys for land registration, and there is as yet nothing like a topographical map of the continent on a small scale. The Department of Defence has lately begun to make maps on the one-inch scale for official use, but the general position is backward. The smaller scale general maps are poorly reproduced in black. There are however two excellent new sheets of the International Map.

Maps of the Union of South Africa.

Thanks to the energy of the late Sir David Gill while H.M. Astronomer at the Cape, South Africa is provided with an excellent framework of geodetic survey; a good half-inch survey of the Orange Free State was made at the joint expense of the British and the then Colonial Government some twenty years ago, and a reconnaissance survey of Basutoland and part of the Cape Colony. Under the Government of the Union topographical surveys are now making some progress, but there are as yet no maps of Natal or of the Transvaal which can be called topographical.

South-West Africa. 1/500,000. Surveyor-General, Windhoek, and O.S.

Sheets covering 2° square making six to each sheet of the 1/M map. Compiled from surveys on the Transverse Mercator projection.

Brown form-lines at 100 feet interval. Water blue; roads red; elaborate conventional signs, with explanation in English and Dutch. Appearance of map spoiled by farm boundaries, names, and numbers. Index diagram showing relation to 1/M sheets. Interesting as the first map on modern lines produced by a South African department.

(Sheet South F–33/6.)

Maps of the Dominion of New Zealand.

The beautiful topography of New Zealand has not received the attention it deserves of cartographers.

New Zealand. 1/125,000.

Drawn at Defence Headquarters, Wellington, and printed in colours at the Government Printing Office.

Relief by contours in red at 100 feet interval, and oblique hill-shading in grey. Railways black. Roads yellow. Water blue. Bush green and too heavy.

The most ambitious topographical map produced in the Southern Hemisphere.

(Sheet Wellington. 1917.)

Maps of British Crown Colonies.

The tropical possessions in Africa are much better surveyed than the older states to the south. Great progress has been made since the establishment in 1905 of the Colonial Survey Committee, whose duty it is to make recommendations for the rapid and economical prosecution of surveys on a systematic plan, and to criticise results in its Annual Report, which should be studied by all students interested in the surveys of the Empire. The Reports of the proceedings at the triennial Conferences of Empire Survey Officers, from 1928, contain also much useful information on Dominion and Colonial Surveys.

In the past the G.S.G.S. have been largely responsible for drawing and publishing the maps of Crown Colonies in Tropical Africa; but Colonial Survey Departments have begun to reproduce their own sheets, superseding the G.S.G.S. series. Old series on various sheet lines are being replaced by new series based on the sheet lines of the International Map. The older sheets are shown in the Catalogue and Indexes of the G.S.G.S. For newer sheets reference must be made to the Indexes of the Survey Departments of the Anglo-Egyptian Sudan (Khartoum), Gold Coast, Nigeria, Northern Rhodesia, Southern Rhodesia, Sierra Leone, and Uganda.

Africa. 1/250,000. Geographical Section, General Staff, and Colonial Survey Departments.

Sheets covering 1° square arranged and numbered as subdivisions of the International Map sheets.

Brown contours or form lines at 100 or 200 feet interval. Water blue; main roads filled yellow.

Some sheets from accurate surveys; others compiled from various sources. Kenya sheets have elaborately classified symbols for vegetation; Uganda less elaborate but cultivation in yellow. Northern Rhodesia sheets distinguish forest from scrub by different green symbols: the Basutoland map, of an older series, is a good example of relief by carefully drawn form lines. Many sheets of the 1/250,000 series are still rough compilations in black, e.g. Anglo-Egyptian Sudan.

Africa. 1/125,000. Geographical Section, General Staff, and Colonial Survey Departments.

Sheets covering 30′ by 30′ arranged and numbered as subdivisions of the above, in various styles and degrees of completeness.

East Africa Protectorate sheets of about 1909 have beautiful form lines at 100 feet interval.

(Sheet South A 37/G II: Nairobi.)

Orange Free State sheets surveyed by the Colonial Survey Section about 1910 have red contours at 100 feet.

Northern Nigeria largely rough compilation with conventional form lines for relief; Gold Coast have form lines at 100 feet interval and layer coloured, with most elaborate conventional signs, but sheets of various sizes, not fitting into the regular system.

(Sheet North B 30/K and part of E. Kumasi. 1925.)

Ceylon. 1/63,360. Survey Department.

Sheets 27 by 17½ inches.

Relief by brown contours at 100 feet interval, every fifth strengthened, and numerous spot heights in brown. Roads red and yellow, towns black, water blue, forest yellow, and various cultivations distinguished by colours.

A good clean map covering the north and east of the island, and superseding the older quarter sheets of similar style now under revision.

(Sheet M 8, 9, 13, 14: Passara. No date, but recent.)

The Department publishes also relief models of the island on the scale of eight miles to the inch: perhaps the only survey department in the World to do such a thing.

Federated Malay States. 1/63,360. Survey Department, F.M.S.

Sheets about 17 by 17 inches covering 15′ square, on the same sheet lines as the maps of the Survey of India.

Relief by brown contours at 50 feet interval, each fifth strengthened, with brown rock-drawing and hill-shading from north-west. Metalled roads filled yellow; water blue; elaborate symbols for types of cultivation. Buildings permanent and temporary distinguished by dots of red and black respectively. Names typed, rather small for scale.

(Sheet No. 2 N/5: Ipoh. 1923.)

Chapter IV

INTERNATIONAL MAPS

The International Map on the scale of 1/1,000,000.
(*Carte internationale du Monde au Millionième.*)

THE publication of a map of the whole world on the scale of one-in-a-million was proposed by Professor Penck at the International Geographical Congress which assembled at Berne in 1891, and at the next meeting held in London in 1895 the proposal was discussed in detail, many resolutions were voted, and a scheme for the division of the sheets was adopted. The Greenwich meridian was chosen as primary, but the question of units—metric or non-metric—was left undecided, the British delegates at that time refusing to have anything to do with the metric system on British sheets.

Little more was heard of the scheme for thirteen years, until it was revived at the Congress held in 1908 at Geneva. But the resolutions then taken must have remained as inoperative as those before, since they bound no one who was in a position to produce a single sheet of the proposed map. Realising that nothing would be done until the project was established on an official basis, the British Government issued invitations to all countries represented by Ambassadors at the Court of St James to send delegates to a Committee which should prepare a definite scheme. This Committee met in London on 19 November 1909, and speedily arrived unanimously at an agreement which was ratified by all the nations concerned. The following is a short summary of the decisions.

1. A uniform set of symbols and conventional signs was adopted, and a characteristic sheet prepared.

2. Each sheet covers an area 4 degrees in latitude by 6 degrees in longitude; but nearer the poles than latitude 60°, two or more sheets of the same zone may be joined. The limiting meridians are reckoned from Greenwich, and the limiting parallels from the equator.

3. The sheets are numbered according to a diagram attached to the Report on a plan which is somewhat complicated. Each sheet has a number, such as

North K. 35.

The letter K signifies the zone of latitude 40° to 44°, the eleventh zone from the equator, as K is the eleventh letter of the alphabet. The number 35 signifies the 35th lune of longitude, each of six degrees, reckoned eastward from the meridian opposite that of Greenwich. Thus a short calculation leads to the result that sheet North K. 35 is the sheet covering Latitude N. 40° to 44° and Longitude E. 24° to 30°. The merits of this system of numbering are not obvious.

Each sheet has also a name, and in practice the sheets are known by their names and not by their numbers, though the latter are important from their place in the system of numbering sheets on larger scales. (See page 57.)

4. The projection is a slightly modified form of the polyconic projection proposed by one of the French delegates, M. Lallemand. It has all the properties necessary for such a map; that neighbouring sheets shall fit along their edges; that the representation of distances and bearings within the sheet shall be sensibly perfect; and that it shall be constructed with ease.

The upper and lower parallels are constructed as in the ordinary polyconic projection, but they are brought slightly closer together, so that the meridians 2° from the centre are their true lengths, instead of the central meridian; and the meridians are drawn as straight lines instead of being slightly curved. These refinements on the ordinary polyconic projection are practically scarcely detectable on the printed sheet, but are theoretically elegant. The tables necessary for the construction of the sheets are given in the Report; they occupy only two pages.

It should be noted that the tables as originally published had a small error of theory which made the numbers tabulated incorrect, though the error was inappreciable in plotting. Correction pages were soon published, but are liable to be overlooked, as they were by the writer, who, after pointing out the mistake, himself printed the uncorrected values in his book on Map Projections.

The tables are calculated on Clarke's figure of 1880. The dimensions of the sheets calculated on any other reasonable figure would not differ by more than a few hundredths of a millimetre.

5. Contours are drawn at vertical intervals of 100 metres, "but in very hilly districts the contours may be at larger vertical intervals, provided that they are spaced at 200, 500, or 1000 metre intervals". The map is coloured on the layer system, according to a scheme attached to the Report.

This colour scale or "gamme" went up into an ugly magenta, but had the great merit that the transition from the green to the brown was through yellow.

6. Minor features of importance, which would not be shown by the contouring, may be represented by shading, but not by hachuring; and

the method of lighting which is most effective for the district may be selected.

7. Precise rules are laid down for the spelling and transliteration of place names. The basis of these rules is that the spelling of every place name in an independent country or self-governing dominion shall be that adopted by the country or dominion. This is of great importance, since it abolishes the customary corruptions of place names as used by foreigners in such countries as Turkey.

Chinese names are to be spelled as in the Chinese Postal Guide. This is a sound rule of practice, but displeasing to those who know Chinese, since the Postal Guide is not consistent in its transliteration, and even transliterates identical names differently, for the sake of distinction.

Although the sheets are intended for international use, they must inevitably show the country of origin in the conventional names written in brackets after the correct forms, and in the pronunciation tables which were to be provided for the more difficult languages.

8. Water features and glaciers are in blue; contours in brown, for the land, and in blue for the sea bottom; roads are in red, and railways in black.

9. Heights above mean sea level are in metres, mean sea level being deduced in each country from tidal observations on its own coasts.

These were the principal resolutions in the report of the Committee of 1909. The first few sheets published showed considerable deviations from the strict rules. Thus in the Highlands sheet produced by the Ordnance Survey the contour interval was altered and the layer tints deranged. The Paris sheet did not follow the colour scheme closely, and the compilation of adjacent parts of Holland and Belgium was defective. The sheet of Kenhardt, Cape Colony, published by the G.S.G.S., followed the rules closely, and showed the failure of the layer system on high plateaus. The sheet of Istambul by the same diverged from the type in its contours and layer colours.

A second Conference met in Paris in December 1913 on the invitation of the French Government. It was agreed that the fundamental resolutions of 1909 on scale, sheet lines, projection, etc. should be considered as definitive, but that the conventional signs, contours, layer colours, lettering, and transliteration were open to revision. The more important changes were as follows:

The normal vertical interval of contours shall be 100 metres. In flat country auxiliary contours may be interpolated, but at regular vertical intervals of 10, 20, or 50 metres. In mountainous country, where it is impossible to contour at 100 metre interval, at least certain principal contours (*courbes maîtresses*) at 200, 500, 1000, 1500, 2000, 2500, 3000, 4000 metres shall be shown, and it is desirable to show the 100 metre contour whenever possible. The principal contours should be in a continuous black line and "auxiliary contours [meaning here the other normal contours] in a dotted line so drawn as to appear like a fine continuous line". The layer colours should follow the scale (*gamme officielle*) attached to the resolutions, and the colour should be changed at each principal contour, but with the following exceptions: Whenever the 100 metre contour can be drawn, the zone 0–200 shall have two shades of green; whenever the 300 and 400 contours can be drawn, the zone 200–500 shall have three shades of yellow (actually buff in the gamme); and whenever the 700 contour can be drawn, the zone 500–1000 shall have two shades of the colour appropriated to it. When these intermediate contours cannot be drawn, the whole zone shall have the tint appropriated to the higher division.

These rather complicated provisions *par exception* reflect the difficulties of obtaining international agreement on a technical subject. The colour scale attached to the official conventional signs sheet was not quite in accord with these rules. The separate *gamme officielle* made scarcely any difference between the two greens. A subsequent "Illustration of the method of construction of the scale of colours. . ." printed by the G.S.G.S. for the *Bureau Central* established at Southampton differed notably in the green tints and the blue. This variety of official pattern is perhaps slightly, but probably not largely, responsible for the variation in published sheets.

The Paris Conference decided to meet again in Berlin in 1914, but instead there came the great European war, and all regular progress was stopped. Only three sheets of Europe had been published, and a few more seen in proof, when there was an urgent demand for a general map of Europe and the Near East on this scale. A staff, largely volunteer, recruited at the Royal Geographical Society, compiled during the war about 100 sheets, which were reproduced partly by the Ordnance Survey and partly by the G.S.G.S.: the former with typed names, the latter carefully drawn. This series followed pretty closely the general lines of the Paris resolutions, and most of the sheets were eventually layer coloured. They show many signs of hasty compilation, but they played a useful part in a crisis, they made the only general map available for the discussions of the Peace Congress and the illustration of the Treaties that followed, and most of them have not yet been replaced by more legitimate sheets.

The progress of the 1/M map since 1919 has been disappointing. Europe is not yet covered by official sheets; the United States have published only four in all, and these mostly sea; and there are very few new sheets for the British Dominions, excepting India, which is complete. The Bureau of the International Map, established at the Ordnance Survey, in its annual report, remarks on the tendency to produce provisional and irregular sheets. But the most serious and rather unforeseen difficulty is that when the sheets are published it is exceedingly difficult to buy them, since there is no proper selling organisation.

One may suspect that this project, like so many international projects, was incapable of realisation. It was too much to hope that thirty or forty reproduction establishments would succeed in following precisely a pattern with which they were perhaps not enamoured, or that the compilation of several hundred draughtsmen would be consistent in interpreting the rules. The only hope of uniformity would have been in making one establishment responsible for drawing, printing, and selling the maps, based on compilations supplied officially by each country, and carefully criticised at the central establishment.

However, this was probably impossible; and the map is at last making some progress. The following is a brief statement of the position in August 1931.

Between latitudes 72° N. and 56° S. there are 1920 sheets in all of which about 1000 cover more or less land. Up to the end of 1929 the last published Annual Report of the Central Bureau at Southampton showed that only 81 sheets had been published in accordance with the Paris resolutions, of which 33 belonged to Europe, 33 to Asia, 5 to Africa, 5 to North America, 3 to South America, and 2 to Australia. Of these only 5 had been published in 1929, 4 of the 5 by the Survey of India. To August 1931 there were 4 more sheets for Europe, 2 for Africa, 8 for North America, and

2 for Australia. And it should be noted that with an eye to statistics the producers are apt to choose for early publication those sheets that have scarcely any land. Thus eight of the European sheets together do not make more than half a sheet of land; the four sheets published by the United States amount to perhaps 1½; and the two Australian sheets first published did not make together an eighth of a sheet.

In addition to the 89 sheets of the regular series there are 10 partially layer coloured, and 11 provisional sheets without layers included in Index I of the Report, which is limited to sheets that are accounted part of the official series. Index II shows several hundred sheets attributed to "Private Societies, etc." These include 25 of the sheets of the Near East compiled mostly at the R.G.S. during the war and published in the best style with layer colours by the G.S.G.S. and 49 sheets of the Provisional Series of Europe compiled by the R.G.S. and reproduced in various degrees of refinement by the O.S.; 40 sheets of the French "Croquis du Sahara" produced by the Service Géographique de l'Armée and transferred from Index I at their request; 50 layer-coloured sheets of Brazil published by the Club de Engenaria do Rio de Janeiro to celebrate the Centenary of their country; and 16 sheets of South America and the West Indies produced in the best style by the American Geographical Society of New York. The last two series are remarkable examples of private enterprise supplanting official effort.

Few of the sheets produced lately conform strictly to the conventions. Some still have contours in brown. Few have realised the intention that the solid *courbes maîtresses* and the intermediate dotted contours should have nearly the same weight; the former are generally too heavy. Very few have obeyed the instruction that when the 100 metre contours can be shown they should be. And few have succeeded in imitating the delicate scheme of layer colouring, to which there are startling exceptions. Roads, railways, and spot heights are often too heavy; the old international boundary symbol is still used. Province and county names are usually too conspicuous. Compilers deliberately neglect the information, especially the spelling of names, supplied by neighbouring states, so that there are instances of alternative sheets produced in self-defence. Such are the difficulties of an international enterprise.

In a general review of the map, one can but regret that it is tied for ever to the debased lettering of the last century: that there is no use of italic on the black plate; that physical names are in a particularly bad sans serif; that there is no design for the marginal information; and that long explanations must be tri-lingual.

But if not altogether an artistic success, which a design by an international committee can scarcely hope to be, it is nevertheless beginning to be very useful as it becomes more complete, and it has had a great effect in unifying the sheet lines of maps on larger scales, which are increasingly arranged as subdivisions of the international sheet lines, to the general convenience of the world.

The international general aeronautical map.

By an international post-war convention the adhering states undertook to publish air maps covering 18° of longitude by 12° of latitude, each comprising the territory of nine sheets of the 1/M map, on Mercator's projection with the curious scale 3 cm. to 1° of longitude on the equator (1/3,708,758). Relief is shown by a few layer tints in grey over pale green or buff ground tint, and by spot heights to one place of decimals of a kilometre, reckoned always to the tenth above the actual height: thus 1715 metres would be shown as 1·8. Rivers are in blue, woods in green dots, railways in maroon, and roads in bistre, with sea pale blue. There is much technical aeronautical information in black. Owing to the great variation of scale of the projection with latitude, the scale of miles must be drawn for each two degrees, and 1/3·7 M on the equator becomes 1/2·2 M in latitude 54° at the middle of the sheet for Great Britain. The last report showed that Great Britain had produced 8 sheets, France 7, and Poland 1. Perhaps because of the inconvenient change of scale the map is not to be continued.

Chapter V

THE MAPS OF EUROPE

A WELL-CHOSEN series of specimens from the principal series of official maps of Europe exhibits every variety of success and relative failure. The technical quality of many is very high, both in engraving and printing; but they do not show much evidence of interest in the artistic side of map production. The beauty of the impression from the engraved copperplate is now little seen; the unpleasing qualities of surface of paper suited to lithography go far to spoil any map; and the newer processes of offset printing, which give excellent results on "Antique" paper, have been little turned in this direction. Nor is there any evidence in European maps of desire to get away from the commonplace if exquisitely fashioned lettering of the middle of last century, back to the purer styles practised by the great masters of the early seventeenth. Ornament is dead, and not even the border is allowed to be decorative. The use of colour has killed the beautiful black of the old maps, without giving them any beauty of colour, for the colours are too often sad. The new French 1/50,000 (not the newest) and some of the recent Scandinavian maps are cheerful exceptions, and the Bulgarian 1/50,000 is commendable for its gay colour.

In the representation of relief, which is the crux of the matter, there is great room for improvement. Hachures in black or in dark brown still ruin the legibility of too many maps, and the system of layer tints has not been very widely used. The Swiss have almost a monopoly of beautiful rock-drawing; the French have been compelled to abandon their excellent double hill-shading in vertical bistre and oblique purple, for it was too expensive. The most interesting experiments in hill-shading are at present on the Norwegian and Danish sheets.

There is a regrettable tendency to minimise the division of the margin in latitude and longitude, and to abolish the graticule; if only for its decorative effect a chequered margin is worth having. By no means all countries are willing to show the relation of their own systems of longitude to the meridian of Greenwich, and there is still in Central Europe a singular affection for longitudes from the reputed Ferro, which is really Paris shifted twenty degrees west.

Since there is not, to the author's knowledge, any guide to the principal series of official maps, other than a very old and out-of-date War Office list, the following brief analysis of the best maps has been made more extensive than would otherwise be suitable for a textbook. Map analysis is an excellent instructive exercise for the student, who might find a wide and almost uncultivated field of research in studying the relative legibility of lettering, the rules of grading in size, and the minutiae which divide success from failure.

The maps chosen for analysis here are from the newer and more important series; many old series have been omitted, as also series on scales larger than 1/40,000. The number and name of the sheet analysed are given in brackets at the end, to serve as a guide to purchase of specimens. It has not seemed desirable in general to indicate the number of sheets published at the date of writing, since some offices publish so much faster than others.

AUSTRIA. 1/50,000. Kartographisches Institut, Wien.

Size 15 by 22 inches.

Relief by contours in brown at 20 metres interval, every fifth strengthened. Lake-bottom contours in blue at same interval, with intermediate contours at 10 metres. Spot heights in black, numerous. Cliff-drawing and vertical shading in brown. Streams blue; lakes pale blue solid tint. Woods in black symbol. Roads, railways, and houses in black. Margins divided in the centesimal system with longitudes from Ferro, and in the sexagesimal from Greenwich: no graticule. Scales of metres and Schritte: 4 Schritte = 3 metres. Names all in black; towns and villages in graded sloping lower case; river names sloping backwards; hill names in engrossing.

A good clear map except in the mountains, where the cliff-drawing is heavy in colour. No characteristic sheet attached. A few sheets only yet published.

(Sheet 4851 West: Attersee. 1924.)

Austria. 1/75,000. Kartographisches, früher Militärgeographisches Institut in Wien.

Size 20 by 14½ inches.

Relief by contours in brown at 50 metres interval, alternate contours strengthened, and vertical hill-shading in brown, with many spot heights. Railways and roads black, water blue; all names black, mostly in graded italic, but hill names engrossing. Forests by small black circles. Graticule not carried across; longitude from Ferro.

The successor to the old 1/75,000 of the Austro-Hungarian Monarchy, and still in provisional form.

(Sheet: Provisional Edition 4956–Neunkirchen und Aspang. 1923.)

Austria. 1/200,000. General Staff Map of Central Europe. K. und K. Militär-Geographisches Institut.

Sheet 1° by 1° (about 15 by 22 inches).

Relief by pale brown vertical hill-shading, and darker brown contours at 100 metres interval. Roads and railways black. Water, but not water-names, blue. Woods tinted pale green. Margin divided in latitude and longitude from Ferro. Sheet numbers are latitude and longitude of centre of sheet. Scale of kilometres and Schritte. (1000 Schritte = 0·75 km.) No adequate characteristic sheet.

A very ugly map which before the war had a reputation not altogether deserved, since during the war it was found that many of the outlying sheets were very inaccurate. Begun in 1889 and about 280 sheets published by 1914. After the war many of the plates passed into the hands of the Succession States, and are reprinted by Czechoslovakia, Yugoslavia, Hungary, Greece, and Romania with new name plates.

(Sheet 41, 41: Saloniki. 1903, revised 1909.)

Austria. 1/300,000. General-Karte von Central-Europa. K. und K. Militär-Geographisches Institut before the war; no longer published.

Rectangular sheets 19 by 16½ inches, covering from Manchester to Constantinople, originally published in black only in 207 sheets in 1873–76, and in colour from about 1881.

Relief by hachures in brown and few spot heights. Detail in black; woods tinted green. Degree meridians but not always parallels carried across sheet which is on a conical projection with central meridian of Paris, from which longitudes are numbered—a very curious feature of the map, especially since on the Paris sheet the central meridian is skewed 11° to the margins.

Austria. 1/750,000. Uebersichtskarte von Mittel-Europa.

Relief by contours and layers. No hachures or hill-shading. Contours at 150, 300, 500, 700 and thence at 300 metres interval; layer tints in ascending shades of brown, rather heavy. Contours in darker brown are the boundaries of the layers; no intermediate contours. "Thalsohlen und Thalebenen," apparently plains in the bottoms of valleys, in two shades of greenish blue. Roads to two principal grades in red; third grade in black; tracks in black, long-dotted. Railways in black. Water, including names of water, in blue. Woods not shown. Mountain names in engrossing. Scale of kilometres. Latitudes and longitudes on margin. Origin of longitudes: Ferro. Meridians and parallels drawn across sheet. No characteristic sheet.

(Sheet H 8: Skoplje. 1901.)

Only six sheets published in this style; others had hachures in brown and no contours; and there were two series, an earlier on Bonne's projection published 1883–86 in 54 sheets, and a later on Albers', of which about 12 sheets published to 1914. A confusing map.

BELGIUM. 1/40,000. Institut Cartographique Militaire.

Sheet 28 by 20 inches.

Contoured in black at 5 metre interval; roads in black filled yellow; woods black tree sign and green tint. Towns black. Rivers blue on black.

(Sheet 53: Dinant. 1923.)

Belgium. 1/100,000. Institut Cartographique Militaire.

Sheet 19 by 16 inches.

Contoured in brown at 10 metre interval; roads in red; woods green symbol; towns black; rivers blue.

(Sheet: Bruxelles. 1922.)

Belgium. 1/200,000. Institut Cartographique Militaire.

Sheet 19 by 24 inches.

Relief by brown contours at 20 metres, very faint and lost in heavy red roads and extensive forest symbol in green. Frequent spot heights. Water and water names blue. Town signs black well graded. Names stamped and ugly. Boundaries by grey ribbon. Horizontal detail but not relief extended into adjoining territory. Not a good map.

(Sheet V: Mons, Charleroi, Dinant to southern frontier. 1930.)

BULGARIA. 1/50,000. Cartographical Institute, Sofiya.

Relief by brown contours at 10 metre interval, every tenth strengthened. Contours not figured except in the margins, and spot heights few. Roads in black, water and water names in blue; woods in green. Villages blocked in vermilion, with names in sloping Cyrillic character; hill names in an ornamental engrossing rather difficult to read. Margins not divided in latitude and longitude.

A well printed and effective map to make about 400 sheets, of which about 30 in the neighbourhood of Sofiya had been published before the war.

(Sheet XII 4: Izvor. 1908.)

Bulgaria. 1/126,000. (Three versts to the inch.)

Printed in black and brown in the topographical department of the General Staff, St Petersburg, about 1880 in 62 sheets 19 by 17 inches.

Relief by contours at close vertical interval occasionally figured in margin, and spot heights.

Now quite out of date and probably unobtainable, but interesting as better engraved than the Russian maps from the same office.

(Sheet VI 4: Slatitsa. No date.)

Bulgaria. 1/210,000. (Five versts to the inch.) Map of Macedonia. Cartographical Institute, Sofiya.

Sheets 13½ by 12 inches.

Relief by brown contours at 50 metre intervals, not figured; many spot heights. Roads in black; water and its names in blue. Place names in Cyrillic, mostly sloping; hill names in ornamental.

A good clear map in about 70 sheets, nearly or quite complete. There is an older series of the same map produced at St Petersburg in the 80's much more finely engraved.

Longitudes from Paris, and projection probably centred on Paris, for the graticule is very oblique to sheet lines.

(Sheet Shkodra. Dated 1902.)

CZECHOSLOVAKIA. 1/75,000. Military Geographical Institute, Praha.

Sheets about 18½ by 15 inches covering 30′ of longitude by 15′ of latitude: longitudes from Ferro. Evidently related to the old Austro-Hungarian map on the same scale and with the same sheet lines, but with revision of detail, and names all written in Czech.

Relief by contours at 100 metre vertical interval and hachures: the whole map in black, except for light green tint on the woods.

(Sheet 4052: Hořovice a Beroun. 1925.)

Czechoslovakia. 1/200,000. Military Geographical Institute, Praha.

This is essentially the Czechoslovak portion of the old Austrian General Staff map with new name plate giving the names in Czech, and with corrections of detail to date. A provisional edition without relief, and a later with the hill-shading in brown of the Austrian map.

(Sheet 36° 49°: Trenčin. 1922.)

Czechoslovakia. 1/750,000. Military Geographical Institute, Praha.

Sheets about 15 by 13 inches, much on the lines of the old Austrian map of the same scale, with detail in black, rivers in blue, and boundaries in red ribbons.

(Sheet C2: Brno, Opava, Nitra. 1919.)

Czechoslovakia. 1/1,500,000. Military Geographical Institute, Praha.

General map of the Republic in one sheet 24 × 11 inches. Layer coloured in yellow, green and brown, with vertical hill-shading in grey. Rivers black, railways red. Subsidiary maps showing population, products, etc. An excellent map dated 1925.

DENMARK. 1/40,000. Generalstabens Topografiske Kort. Generalstabens topografiske Afdeling, Kjöbenhavn.

Sheet 18½ by 15 inches.

Relief by contours in black at intervals of 2 metres, every fifth pecked. Numerous spot heights. Streams black overprinted blue; lakes ruled blue. Woods by tree signs in black, and brown tint. Roads in black, filled yellow. Railways and houses in black. Margins divided at wide intervals; longitudes from Copenhagen. No graticule. Scales of Alen and metres. Small characteristic sheet.

A clear well-drawn map rather spoiled by the brown woods. A new series begun in 1907, superseding the old in hachured black.

(Sheet Odense. Dated 1920.)

Denmark. 1/100,000. (New.) Generalstabens Topografiske Afdeling, Köbenhavn.

Sheet 16¼ by 13 inches.

Relief by contours in brown at 5 metre interval, every fifth pecked. Many spot heights in black. Roads black (principal filled brown), town plans black, railways black. Streams black overprinted blue, sea lined blue with black dotted contours at 5 metres. Names all in black, mostly italic and stump, but physical names sloped backwards.

Graticule carried across; longitudes from Copenhagen. Very elaborate characteristic sheet.

Very well drawn and printed map begun in 1919, and about 70 of the 100 sheets already published; a great improvement on the former uncontoured 1/100,000.

(Sheet No. 44: Kolding. 1921.)

Denmark. 1/160,000. Generalstabens Topografiske Kort.

Heliogravure in black only.

Relief by contours at 5 metres vertical interval (remarkably close for so small a scale, and possible only in a flat country). Contours not figured, but spot heights numerous. Principal roads double black line. Railways chequered. Lakes with horizontal water lining, to distinguish from arms of the sea with water lining parallel to shore. Scales of kilometres and Danish miles. Latitudes and longitudes in margin. Origin of longitudes Copenhagen. Meridians and parallels carried across the sheet at intervals of 6′ longitude and 4′ latitude. Sheets rectangular. River names sloping slightly backward.

Full characteristic sheet. A good example of engraving, but suffers from want of colour. Five classes of boundaries; too many for the scale.

(Blad 8. 1910.)

Denmark. 1/160,000. (New.) Generalstabens Topografiske Kort.

Sheet 15½ by 15 inches.

Transfer from the black engraved plate described above, overprinted in colour. Sea pale blue; lakes deeper blue and outlined still deeper. Land pale buff, woods and plantations deeper buff, heath pink, marsh green. The black engraved rivers and streams overprinted with blue line.

There are two black plates, one with contours and one without. Some of the new coloured edition from one and some from the other.

(Blad 14 in three styles, coloured with brown contours, and two in black, one without contours.)

Denmark. 1/200,000. Generalstabens Topografiske Kort.

Rectangular sheets 20 by 15 inches. Longitudes from Copenhagen.

Frequent spot heights; 4 and 10 metre contours in sea dotted black. Roads brown. Water blue lined and narrow solid ribbon. Very elaborate conventional signs. Woods brown, marsh green, heath pink. Every tenth kilometre post numbered on roads. A good clear map.

(Sheet 15: Maribo. 1928.)

Denmark. 1/1,000,000.

In one sheet in the style of the International Map but not on its sheetlines.

FRANCE. 1/50,000. (Type 1900.) Service Géographique de l'Armée.

Sheet 21 by 15 inches.

Relief: contours and double system of hill-shading. Contours in brown at 10 metre vertical interval, but not figured. Spot heights frequent. Vertical hill-shade in bistre; oblique shade from north-west in purplish grey. Roads in black, four grades; tracks in black, two grades. Buildings in red. Water blue. Woods, meadows, and gardens in different shades of green; orchards and vineyards in purple. Population of each village in red.

Scale of kilometres. Latitudes and longitudes shown on margin in the ordinary and the centesimal division. Origin of longitudes Paris. Meridians and parallels not carried across the sheet.

An excellent feature is that the margins are left open, and the principal roads, railways, etc. are continued off the sheet, together with anything which would suffer from being cut across by the sheet margin. Writing very plain. Names of water in blue. Elaborate characteristic sheet, including separate signs for factories operated by steam, water power, and electricity respectively.

(Sheet XXII–14: Versailles. No date.)

Sheet XXXVI–33: Tignes, of the same series, on the Alpine frontier, with its contoured glaciers and cliff-drawing, is a splendid example of the series, and one of the finest topographical sheets ever produced.

Sheet XXXV–32: Bourg St Maurice (1927), with contours in orange, is an interesting variant.

France. 1/50,000. Édition provisoire. Service Géographique de l'Armée.

Sheets 20¾ by 15¾ inches.

This simplified but still very effective map has contours in brown at

10 metre interval, with oblique hill-shading in brown from the north-west; woods and cultivation in black symbols and pale green tints; roads black; rivers blue. The writing is all in sans serif, which is regrettable; but it is a fine clear map, and much less expensive to produce than the original 1/50,000.

(Sheet XXXVI–15: Sarrebourg. 1920.)

The "Carte de France au 50,000e. Type 1922" is very like this provisional edition, and is destined to supersede the more beautiful but too expensive style of the earlier 1/50,000. Good examples are sheet Thionville (1924), and Névache (1928) with fine black cliff-drawing and glaciers contoured blue.

France. 1/80,000. Dépôt de la Guerre (now Service Géographique de l'Armée).

The older editions, in large sheets 31 by 19½ inches, printed in black from the engraved plate, the recent rather poorly photolithographed. Heavily hachured but not contoured. A very celebrated map of the old school, with scales in metres, leagues, and toises; latitudes and longitudes in the sexagesimal and centesimal systems, and co-ordinates of the sheet corners from the Panthéon calculated on Bonne's projection.

The British Army went to France in 1914 with sheets of this map overprinted with approximate contours in red. The Germans invaded France with reproductions of the same map overprinted with village populations and indications of fortified areas; neither side had anything like modern colour-printed maps of the battle area at the beginning of the war.

France. 1/80,000. Frontière des Alpes.

A special edition of the 1/80,000 without hachures, but with contours at 25 metres and cliff-drawing in brown, and pretty glacier-drawing, not contours, in blue. Woods in dull green; water blue. Note in margin: "Les courbes de niveau, qui sont régulières dans la partie française, sont fictives dans la partie italienne".

France. 1/100,000. Dressée par ordre du Ministre de l'Intérieur. Published by the Librairie Hachette.

No contours, but soft grey vertical hill-shading. Towns black, roads red, woods green symbol, water blue. Spot heights in blue. Graticule figured in the sexagesimal system, carried across.

(Sheet XXI–26: Lyon (Sud-ouest). 1927.)

France. 1/200,000. Service Géographique de l'Armée.

Sheets 26 × 16 inches, on Bonne's projection.

Relief by contours and vertical hill-shading. Contours at 20 metres vertical interval, very finely drawn, in pale brown, and not figured. Hill-shading brown. Roads and tracks, six grades, in red. Railways in black. Water in blue. Woods in green; tree signs. Meridians and parallels carried across map, inclined to margins, figured in centesimal division of the quadrant. Margin divided also in kilometres from the Panthéon. Full characteristic sheet. The contours drawn so lightly and so much overlaid with names and detail that the relief is illegible.

(Sheet 8: Abbeville. No date.)

France. 1/320,000. Dépôt de la Guerre.

Sheets 32 by 20 inches.

Hachured, no contours, and printed in black from the engraved plate. Similar in style and detail to the old 1/80,000.

France. 1/500,000. Carte de France et des pays limitrophes. Service Géographique de l'Armée.

Sheet 17 by 17 inches.

Relief by brown contours at 100 metres and oblique brown hill-shading. Roads, railways, and town signs black; water and water names blue. Woods green tint. Margin divided in centesimal and sexagesimal systems, longitudes in former from Paris and latter from Greenwich. Four sheets to each sheet of the International Map, numbered accordingly. A good clear map, but names perhaps rather light. Will supersede the old 1/500,000 of the Dépôt des Fortifications uncontoured with vertical hill-shading brown.

Another edition: "Type Aviation" with principal cities and roads in red; rivers stronger in solid blue; no contours, light grey hill-shading, spot heights in large black figures. Conspicuous symbols for aerodromes and "terrains de secours".

(Sheet N.M. 32–S.O.: Strasbourg. 1926.)

GERMANY. 1/50,000. (New Series.) Württemberg, Statistisches Landesamt.

Sheets about 19½ by 17½ inches, covering 20′ longitude by 12′ latitude. Longitudes from Greenwich.

Relief by brown contours at 10 metre interval. Towns and roads black. Marked footpaths distinguished by complicated red symbols. Woods pale green tint. Rivers and water names dull blue. Hill names written in circular arcs. Margins weakly divided in minutes and sparsely figured. Grid of 2 km. squares in the "German co-ordinate system".

A new series of which only a few sheets yet published, to supersede the old 1/50,000 of the Kingdom. The black plate scarcely heavy enough to stand the colour.

(Sheet 1115: Reutlingen. 1925.)

Germany. 1/50,000. Norddeutschland. Reichsamt für Landesaufnahme, Berlin.

The same, except for absence of marked footpaths and green woods. Interesting marginal diagrams of political boundaries, and of compass variation referred to grid north. Longitudes from Greenwich and Ferro.

(Sheet 511: Drossen. 1927.)

Germany. 1/50,000. Bayern. Topographisches Bureau.

Nearly the same; but no woods or marked footpaths; longitudes from Ferro; graticule carried across; no grid, but unexplained division of margin into spaces curiously lettered and numbered.

(Sheet 904: Königshofen. 1923.)

Though differing slightly, all these sheets are headed "Deutsche Karte 1/50,000" and will presumably be gridded in one system, of which the promised characteristic sheet is not yet published.

Bavaria. 1/50,000. (Old Series.) Topographisches Bureau of the K.B. Generalstab.

Sheet 16 by 20 inches.

Relief by contours in brown at 20 metre interval, cliff-drawing in brown, and very heavy oblique hill-shading in purplish grey. Many spot heights. Water in pale blue not nearly strong enough. Roads black. All names black, in italic and stump. Graticule across sheet and margins divided to ten seconds, but very imperfectly figured; origin of longitudes indeterminate. Horizontal scale in metres and geographical miles, of which one equals 7419·55 metres.

A heavy map, difficult to read in the mountains, of which few sheets were published.

(Sheet Scharfreiter. 1907.)

Germany. 1/100,000. Preussische Landesaufnahme.

Sheet 12¾ by 10¾ inches.

Relief by contours in brown at 50 metre interval and brown hachures. Many spot heights in black. Railways, roads, and town plans in black, rivers in blue, rather light, lakes lined blue. Forests in black tree symbol. Graticule not carried across, margin obscurely divided, longitudes from Ferro. All names in black, mostly Roman, but water names italic sloped backwards.

A finely engraved and well-printed map in which detail is rather obscured by excess of hachures in country of small relief. The hachures survive from the older edition which was uncontoured. 674 sheets in all, of which about 300 on the new edition.

(Sheet 132: Christburg. 1920.)

Bavaria. 1/250,000. K. Topograph. Bureau.

Relief by contours, hill-shading, and layers. Contours at 100 metres interval throughout; contours at boundaries of layer tints strengthened; all printed in brown; some intermediate contours in brown long-dotted. Contours not figured, and spot heights infrequent. Hill-shading light grey, oblique, north-west light. Layer colours in spectrum order, leading to very clear red for high mountains. Colours change at 300, 400, 500, 600, 800, 1000, 1200, 1500, 2000, and 2500 metres. Roads in black. Water in blue. Woods not shown. Scale of kilometres. Latitudes and longitudes on margin. Origin of longitudes not stated; evidently Ferro. Meridians not carried across sheet. Trans-frontier country left nearly blank. Writing: place names italic and stump; mountain names sans serif; water names in blue, sloped to left. No characteristic sheet attached.

The general effect of this map is good; but only three sheets were ever published.

(Sheet 8: Munich to southern frontier. 1906.)

Germany. 1/200,000. Topogr. Uebersichtskarte des Deutschen Reichs. Kartogr. Abtheilung der K. Preuss. Landes-Aufnahme.

Sheets about 13 by 11 inches, each covering 1° of longitude by 30′ of latitude. Begun 1899 and nearly complete in 196 sheets. Printed still from engraved copper plates.

Relief by brown contours at 20 metres, each 100 strengthened and 10 dotted in flat country (Edition A). Edition B for hilly country has 50 metre dotted, 100 continuous, and 400 strengthened, according to catalogue. In hilly country Edition A much overcrowded with contours. Few spot heights. Roads and railways in black, rivers blue rather lightly drawn. Lakes horizontal blue lined; valley bottoms tinted green. Woods black symbol. Good characteristic sheet. Graticule not carried across; margin obscurely divided, with longitudes from Ferro.

The lettering of this map has much exaggerated horizontal serifs, and is an excellent example of how illegible such serifs may make an otherwise well-engraved letter.

(Sheet 153: Bamberg. 1910.)

Germany. 1/200,000. Topograph. Specialkarte von Mittel-Europa (formerly known as the Reymann map).

Very finely engraved and printed in black heliogravure. Detail very nearly the same as the German 1/200,000 above, but on smaller sheets 13½ by 9 inches rectangular. Graticule oblique, and names in different style. Begun in 1806, and still published with corrections to date by the Kartogr. Abtheilung der Preuss. Landes-Aufnahme.

Relief by hachures, over-heavy in hilly country. Extends from Eastern France to Southern Scandinavia and to Austria.

Germany. 1/300,000. Uebersichtskarte von Mitteleuropa. Kartogr. Abtheilung der Preuss. Landes-Aufnahme.

Sheets about 18 by 14½ inches, each 2° of longitude by 1° of latitude, covering Central Europe without Austria and Succession States. Begun 1894 and about 120 sheets published.

Relief by vertical hill-shading in brown; no contours. Roads in black filled orange, railways black. Rivers good strong blue. Woods black tree symbol and dull green tint. Lakes blue lined. Lettering clearer than in the 1/200,000 but poor in form.

(Sheet M 52: Halle. 1909.)

The new edition of this map has a ten-kilometre grid in very fine black lines; a chart showing the compass variation from grid-north, and scales of degrees by which the map may be orientated by compass. Grids centred on meridians 3° apart, and numbered so that 3600 of strip 9 adjoins 4400 of strip 12. Co-ordinate card on margin to be cut out and used for measuring co-ordinates on the grid.

(Sheet M 52: Halle. 1909, with corrections 1927.)

A provisional series of sheets covering 1° lat. by 2° long. from Ferro in black with green forest extends eastwards into Poland and Russia. No grid.

(Sheet Y 59: Korjelskoje. 1927.)

GREECE. Khartes Ellenikon Basileion. 1/75,000. K.K. Militär. Geogr. Inst. Wien.

Relief by vertical grey shading. Contours at 20 metre darker grey; 100 metre contours strengthened; occasional intermediate 10 metre contours long-dotted. Sea bottom contoured. Cliff-drawing brown. Principal features across frontier shown, with 100 metre contours and hill-shade. Roads red, in three or more grades. Tracks in red. Water blue. Woods black, with tree signs. Scale of kilometres and Bemata (paces). Scales of slopes corresponding to contour intervals (this an unusual feature). Latitudes and longitudes on margin. Origin of longitudes: Athens, with reduction to Greenwich. Meridians not carried across sheet. Writing in modern Greek character. Complete and elaborate characteristic sheet attached.

Other sheets of series have brown contours and no shading.

(Sheet 40° 1° E: Rapsane-Tempe. 1909. Sheet Atalante. No date.)

Greece. Peloponnesus. 1/100,000. Greek General Staff.

Sheets about 17 by 22 inches covering 30′ of longitude from Athens by 30′ of latitude.

Relief by contours in grey at 40 metres, each fifth strengthened, and few spot heights. Roads red; water blue. Names in Greek. Elaborate symbols.

(Sheet II Δ: Corinth. No date; about 1928.)

Greece. 1/200,000.

Sheets based on the old Austrian 1/200,000, but with each fifth contour strengthened. Names in Greek.

(Sheet 38° 40°, i.e. long. from Ferro and lat. of sheet centre: Kerkyra–Argyrokastrou. 1917.)

Other sheets of the same series without green woods extended northwards.

(Sheet 45° 42°: Purgos (Burgas). No date, about 1927.)

Greece. 1/300,000. K.K. Militär. Geogr. Inst. Wien.

Sheets 17 × 16½ inches.

Detail black. Relief by brown hachures. Longitudes from Paris and Greenwich. Really part of the Austrian 1/300,000, *q.v.*

HOLLAND. 1/50,000. Topographische Inrichting.

Sheets 15¾ by 10 inches; later 19¾ by 15¾.

Contours in black at 10 metre intervals, too great an interval for the flat country; roads black filled orange; woods green tint with black symbol. Towns black; rivers filled blue-black.

(Sheet 45: 's-Hertogenbosch W. 1927.)

Holland. 1/200,000. Ministerie van Oorlog, Topographische Inrichting.

Sheets 15½ by 10 inches.

Printed in black from finely engraved plate. No contours.

(Sheet 13: Biesbosch. 1925.)

ICELAND. 1/50,000.

Sheet 17¼ by 15¾. Printed in colour by the Generalstabens topografiske Afdeling, Kjöbenhavn.

Relief by contours in reddish brown at 20 metre interval, changing to blue on the icefields, with moraines in black. Rocks in black. Streams overprinted blue on black. Lakes and sea lined blue with solid ribbon margin. Green symbols for vegetation. Graticule not carried across; longitudes from Copenhagen. Horizontal scale of metres and alnir; 1000 alnir equal 625 metres.

A beautiful and ambitious map, planned in 115 sheets, now covering south and west coasts.

(Sheet 97: Kalfafelsstadur. 1905.)

Iceland. 1/100,000. Generalstabens Topografiske Kort, Kjöbenhavn.

Sheet 17 by 16 inches.

Relief by brown contours at 20 metres, each fifth weakened by pecking; cliff-drawing in black. Roads, bridle paths, marked and unmarked paths in different black symbols overprinted brown. Water blue. Elaborate surface symbols for homefields, meadow, bog, grass in black with green tint, moraine, lava, sand, etc. in black. Explanation in Icelandic, Danish, and English.

(Sheet 43: Blonduos. 1921.)

Iceland. 1/250,000. Yfirlitskort. Geodetic Institute, Copenhagen and Reykjavik.

Sheet 28 by 20 inches.

Relief by brown contours at 25 metres and black cliffs. Roads brown; meadows green. Elaborate symbols much as in 1/100,000 above. Water blue.

(Sheet Sudvesturland. 1930.)

ITALY. 1/50,000. Istituto Geografico Militare, Firenze.

Printed in black in the '80s on sheets 15 by 14½.

Relief by contours in black at 25 metre interval, and cliff-drawing. Every fourth contour strengthened. Crowded with contours and difficult to read.

Italy. 1/75,000.

In black, similar to the 1/50,000, but with hachures and contours at 50 metres.

Italy. 1/100,000. Grande Carta Topografica del Regna d' Italia. Istituto Geografico Militare, Firenze.

This map is published in many editions: in black with contours and hachures (*tratteggio*) from copper and from zinc; in colour with contours only, with contours and hill-shading, and with contours and tratteggio; in brown with administrative boundaries; and in *calco pallido* (unidentified) each on three different kinds of paper: light (*tipo canape*), stiff to fold; ordinary (*carta commune*); and thick without size (*carta pesante senza colla*), of which the first is normally supplied, not the second. This is from the catalogue of the Istituto. The sheets show great variety not easy to identify with the above descriptions.

Sheets 32: Como (1907); 64: Rovigo (1918); and 211: S. Arcangelo (1927) have contours in brown at 50 metres not figures, with brown oblique hill-shading; roads black, water blue, woods in sparse green symbol. Water names in black.

Sheet 24: Maniago (1912), and many later sheets have contours occasionally figured, hill-shading grey, and a pale grey ground tint. Roads and tracks red, water names blue, woods in green lining with green spots.

Sheet F 5: Val Formazza (1907), has effective vertical hill-shading in buff and oblique in brown. Glaciers contoured blue and ice-falls in blue lining. Moraines black. Lakes solid greenish blue. Woods in small green symbol.

All sheets 30′ by 20′, about 16 by 14 inches.

Italy. 1/200,000. Istituto Geografico Militare.

Sheets 15¼ by 14½ inches, covering 1° of longitude by 40′ of latitude.

Relief by contours and hill-shading. Contours at 100 metre vertical interval. Hill-shade in grey-brown, oblique. First and second grade roads in red; third grade, black in the plains and red on the mountains. Minor roads made and unmade, mule tracks, footpaths easy and difficult, and mountain passes all distinguished by characteristic signs. Railways black. Water blue. Towns by signs to indicate degree of importance. Double, single, narrow gauge, and tramways of various kinds, all distinguished by signs. Woods pale green. Scale of kilometres. Latitudes and longitudes shown on margin. Origin of longitude: Rome, Monte Mario. Meridians and parallels not carried across sheet. International frontier with broad band of colour. Trans-frontier country in full detail. Mountain names in engrossing, difficult to read. Good characteristic sheet attached.

Sheet 11: Brescia. 1908.

Sheet 15: Oulx (1910), has good glaciers in blue line.

Sheet 3: Tirano.

Oblique brown hill-shading on a lighter buff vertical shade, or perhaps a ground tint. Glaciers contoured in blue with ice-drawing. Forests in green tint. Names hard to read in the mountain country, but otherwise a very good map.

Italy. 1/500,000. Istituto Geografico Militare.

Relief by contours at irregular vertical intervals, in brown, alternately continuous and pecked. Contours at 1, 2, 3, 5, 8, 10, 13, 16, 20,00 metres. Layer colouring in green and brown, changing at each contour. Roads and railways black. Rivers and lakes blue. Sea blue lined, with contours in blue. Latitudes and longitudes in margin, each degree carried across map. Longitude from Rome. Scale of kilometres. Writing all in black. Town names in upright and italic capitals, and case; villages in stump. River names in stump. Hill names in engrossing sloped slightly backwards. Rail- and tram-ways in three grades; roads in four. Mule tracks, footpaths, and "difficult" paths distinguished. A good map coloured rather heavily.

Sheet 24: Napoli. 1914.

Sheet No. 9: Udine (1914), has same contours as Napoli, continued by 400 metre intervals above. Layer colours going from brown through olive-green to a blue like the lakes. Not very successful.

NORWAY. 1/100,000. Topografisk kart over kongeriket Norge.

Sheet about 22 by 15 inches covering 1° of longitude by 20′ of latitude.

Relief by contours in black at 30 metre vertical interval. Contours not figured, except at margin, but spot heights numerous. Principal roads double black line. Lakes blue. Rivers black. Railways black. Scale of kilometres and "geografisk mil" (1 g.m. = about 7·4 km.). Latitudes and longitudes on margin. Origin of longitudes, Kristiania. Meridians and parallels not carried across the sheet. Writing: town names in case, villages in stump; water names in black, sloped to left; mountain names in engrossing. Full characteristic sheet.

Sheet: Tunhovd. 1920.

Sheet F 34: Kröderen (1920), of this map has brown vertical hill-shading, rather heavy, making it hard to read.

Sheet Harstad (1912).

Differs from above by showing relief in layers of pale green and brown, with oblique hill-shading from west in purple. Contours brown (except sea level, blue). Rivers blue. A clear and effective map. Another edition of the same sheet without layer colouring or hill-shading is not so effective, but easier to read detail.

Norway. 1/250,000. Landgeneralkart over Norge. Norges Geografiske Opmåling.

Sheet 21½ by 18 inches.

Relief by contours in brown at 50 metre interval, faint and vague layer tints in pale green and buff, and oblique hill-shading from the west in greyish purple. Contours on glaciers dotted blue. Frequent spot heights, but contours not figured, except on margin. Rivers blue; lakes solid pale blue. Roads, tracks, and railways in black. Margins divided to five minutes, and graticule carried across the map. Longitude from Kristiania. Scale of kilometres and "geographical miles" of which one equals about 7·4 kilometres.

A well drawn and very cleanly printed map: a good model for the representation of mountains and glaciers.

(Sheet XXIV: Rana. 1921.)

POLAND. 1/100,000. Geographical Institute.

Sheets covering 30′ of longitude by 15′ of latitude. A series based on the German map of West Russia made and published by the General Staff during the war from the Russian 1/80,000 which was contoured in *sajen* of 7 feet. The German map reproduced these contours and figured them to the nearest metre, giving intervals 4 or 5 to the contours at 2 sajen interval. The Polish map mostly reproduces these contours, but some sheets are at 10 metre intervals. Water blue, a great improvement on the German black; woods green tint or black symbol. Longitudes mostly from Ferro but some from Pulkovo. Very well drawn and reproduced for an evidently provisional series.

(Sheet A 36, B 42: Baranowicze, 1923, for comparison with German S 29. Or sheet A 49, B 41: Załosce, 1925, for metric contours.)

PORTUGAL. 1/50,000. Direcção Geral dos Trabalhos Geodesicos e Topograficos, Lisboa.

Relief by contours in red at interval 25 metres and frequent spot heights. Railways black. Roads red. Rivers blue. Town and village names black in lower case; river names (very infrequent) in blue; woods in green tree signs. Meridians and parallels at each 5′ carried across sheet. Sheet rectangular. Double division of longitudes, from Greenwich and Lisbon. Many conventional signs and full characteristic sheet. A good clear map.

(Sheet No. 11 *d*: 1919.)

Portugal. 1/200,000. Direcção Geral dos Trabalhos Geodesicos e Topograficos, Lisboa.

Sheet 31 by 20 inches.

Relief by brown contours at 50 metres unfigured and many spot heights in black. Roads red; water and water names blue; woods green symbol. Scale of kilometres and *leguas* of 5 km. Little more than a road map.

(Sheet 7, including Lisbon. 1915.)

ROMANIA. 1/50,000. Institutul Geografic, Stat. Majorul General al Armatei.

Sheet 15¾ by 16 inches.

Relief by contours in brown at 10 metre interval, figured only on the margins, each tenth strengthened. Many spot heights in black. Roads in red and in black. Streams in blue, but all names in black, very large and overpowering the detail. Orchards dark green, woods light green. No graticule.

A well printed but sad coloured series of about 200 sheets, of dates 1891 to 1911.

(Sheet Scheia, No. XVI S. 1899.)

Romania. 1/100,000. Institutul Geografic al Armatei.

Sheet 15½ by 16¾ inches.

Relief by contours in brown at 20 metre interval, strengthened at each 200, with supplementary at intervals of 10 and 5 in the plain; only the latter figured, except in the margins; many spot heights. Town plans emphasised by red spots; roads in black, with principal filled red and yellow. Water blue. Names all in black. Coloured in tints to show cultivation and vegetation.

Well and clearly engraved in about 100 sheets dated 1906 to 1913. Graticule carried across sheet. Longitudes from Paris.

(Sheet Focsanii, No. IX K. 1909.)

Romania. 1/200,000.

A Romanian edition of the old Austrian 1/200,000, printed apparently from transfers of the old plates, with names translated into Romanian.

RUSSIA. 1/100,000. Chief Geodetic Committee of the V.S.N.KH. of the U.S.S.R. Moscow.

Sheets 20′ by 20′, 144 to each sheet of the International Map, and numbered accordingly.

Relief by brown contours at 10 metres with intermediate pecked contours, and spot heights to 0·1 metre in black. Contours figured and each tenth strengthened. Roads and town plans black. Water and spot heights along streams blue. Woods green lined. Elaborate characteristic sheet and long explanations on covering sheet attached.

Larger scale sheets on 1/50,000 and in places on 1/25,000. Smaller scale series on 1/200,000 and 1/500,000 projected.

(Sheet O 36–105: Kuvshinovo or Sheet L 37–38: Berdyansk.)

Russia. 1/126,000. (Three versts to the inch.) General Staff, St Petersburg.

Sheet 23 by 16½ inches, covering western European Russia only, in about 500 sheets. Printed in black only, and hachured. Graticule carried across sheet, longitudes doubled, from Pulkovo and Paris. Superseded by the above.

Russia. 1/420,000. (Ten versts to the inch.) St Petersburg, Topographical Section of the General Staff.

Sheet 25 by 19½ inches, published between 1868 and 1889 with revisions. About 150 sheets.

Relief by slight hachures in brown and spot heights in sajen. Roads and railways black, water blue, sea and lakes solid blue with waterlining. Forests green. Crowded with small names very hard to read in the Russian character. Graticule carried across sheets, becoming very skew when far east or west of Pulkovo, the origin of longitudes.

Caucasus. 1/42,000. (One verst to the inch.) Caucasian Military Staff, Tiflis (in Russian).

Sheets 19 by 20½ inches.

Contours black at 10 sajen or 70 feet, each tenth strengthened; black rock-drawing; glacier contours and ice-drawing blue; water blue. A rough but effective map of the Western Caucasus in many hundred sheets, always very difficult to obtain and presumably to be superseded, with other Caucasus maps.

Caucasus. 1/84,000. (Two versts to the inch.)

Sheets same size, each covering four of the above.

Black contours at 10 sajen and cliff-drawing. Water blue. Covers same area and extends south. Hard to obtain.

Caucasus. 1/210,000. (Five versts to the inch.)

Sheets about 24 by 18 inches, covering the whole Caucasus.

Brown hill drawing and black spot heights; glaciers lined blue. Roads and water black; forests green tint.

SPAIN. 1/50,000. Dirección General del Instituto Geografico y Estadistico.

Relief by contours in brown at 20 metre interval, figured in black rather confusedly. Town plans and main roads in red; rivers blue; forest, orchard, pasture, in separate green symbols. Rest of sheet covered with fine parallel wavy black lines, making small names indistinct.

Sheet covers 20′ of longitude by 10′ of latitude, bounded by meridians and parallels, not carried across. Margin fully divided. Longitudes from Madrid. Scale of kilometres. No characteristic sheet attached.

(Sheet 1036: Olvera. 1918.)

Spain. Mapa Militar. 1/100,000. Depósito de la Guerra.

Relief by contours in orange, not figured, at interval 50 metres, and rudimentary hill-shading in black. Railways and roads in black. Rivers blue. Forests small green symbol. All names in black. Town plans black. Margins divided in the centesimal and sexagesimal systems; sheets cover 60′ long. by 40′ lat. centesimal; each 10′ carried across.

An early sheet (in Mallorca) of a new series intended to cover all Spain.

(Sheet 214: Inca. No date [about 1915].)

A sheet of the same series (86 and 62) Seo de Urgel, dated 1920, on the mainland, has contours in brown, roads and town plans in red, and no hill-shading. Perhaps more legible, but not so pleasing in appearance as the earlier sheets of Mallorca.

SWEDEN. 1/200,000. Generalstabens Litografiska Anstalt, Stockholm.

Relief by hachures below about 550 metres and contours above; both black. Hachures vertical; contours shaded to bring up relief, apparently arbitrarily. Contours not figured, and spot heights insufficient to show contour interval. Combination of hachures and contours in this way unusual. Principal roads double black line, yellow between. Lakes blue. Rivers black. Scale of kilometres. Latitudes and longitudes on margin. Origin of longitudes: Stockholm, with reduction to Greenwich. Meridians but not parallels carried across sheet. Woods not shown very clearly; small black tree signs on the hachures hardly distinguishable. Trans-frontier country left entirely blank. Writing: no distinction between hill names and villages; names of lakes sloped to the left. No characteristic sheet attached.

(Sheet 58: Kolasen. 1905.)

Sweden. 1/400,000. Generalstabens översiktskarta över Sverige.

Sheets 19 by 16 inches with graticule; longitudes from Stockholm.

Relief by vague hill-shading in brown, and spot heights in black; no contours. Roads in two classes of brown, much confused by administrative boundaries in black overprinted orange. Water blue; woods a heavy yellowish grey. Water names sloped back. Map confused by number of lakes.

(Sheet VII: Norrkoping. 1913.)

Sweden. 1/500,000. Generalstabens Litografiska Anstalt, Stockholm (Albert Bonnier's Förlag).

Two series in sheets 19 by 14 and 14 by 9½ inches. In the former relief by contours in black at 100 metres, with dotted subsidiaries at 33 and 66 up to 500; layer colours in yellow and brown changing at each hundred. In the latter brown contours at 200 metres with intervening hundreds dotted, and layers green, buff, and brown changing at 200. The former scale turns to grey at the top, the latter to purple. Roads and rivers black in both. Water names black sloping back. In former lakes blue lined with marginal solid ribbon, in latter solid blue. Clear well printed map without too much detail.

(Sheet VI of first series, 1915.)

Sheet 25: Byske-Boden of second series; no date, but about 1915.

SWITZERLAND. 1/25,000. Eidg. Topograph. Bureau, Bern.

Relief by contours in brown, at 10 metre vertical interval; every tenth contour long-dotted and figured. The contours are so close that they are very hard to follow. Cliff-drawing in black. Roads in black, of two grades.

Tracks in black, long-dotted. Scale of kilometres, and map divided into squares of approximately 6 cm. Latitudes and longitudes shown on margin. Origin of longitudes not stated, apparently Paris. Meridians and parallels not carried across the sheet. Woods shown by very minute tree signs in black, making the pale brown contours still more difficult to read. Water blue. Writing in italic; no distinction between physical features and village names. No characteristic sheet attached.

(Sheet 376: Pilatus. 1894, revised 1906.)

This series covers the northern half of Switzerland: the series on 1/50,000 covers the southern. The two series make what is called the Siegfried Atlas.

Switzerland. 1/50,000. Eidg. Stabsbureau.

Relief by contours at intervals of 30 metres, in brown on ground covered with soil and vegetation, in black on bare rock or moraine, and in blue on glaciers. Cliff-drawing in black. Margins divided curiously: first in longitude east of Paris, and in latitude; secondly in kilometres, numbered increasing east and north, with Berne Observatory (presumably) given the arbitrary co-ordinates 600, 200. Further, the edges of the rectangular sheet are labelled with their distance from Berne in metres, so that the E.—W. line numbered 146 bears the label 54000^{m}. S., while that numbered 134 is labelled 66000^{m}. S. Note that $146 + 54 = 134 + 66 = 200$. Similarly for the N.—S. lines, which in this sheet are labelled 2500^{m} and 20000^{m} east of Berne, and numbered 602·5 and 620 respectively. The sheet is further divided into squares of 3 km. forming part of a continuous system from Berne, and not fitting the horizontal dimensions of the sheet which covers 17·5 km. wide. The apparent confusion of the co-ordinate system has doubtless an historical explanation. In all other respects the sheet is admirable. The contours of this map are now being revised by stereographic survey.

(Sheet 473: Gemini. 1918.)

Switzerland. 1/50,000. Schweizerische Landestopographie, Bern.

The ordinary 1/50,000 combined into large sheets and overprinted with vaguely defined layers of pale green and buff, with rather strong oblique purple hill-shading. Pictorial and effective rather than systematic.

The 3 km. squares survive but the margins are not figured either in lat. and long. or in rectangular co-ordinates.

(Combined sheet: Vierwaldstättersee. Edition of 1907.)

Switzerland. 1/100,000.

Printed in black from engraved plates.

Relief shown by hachures, darkened on the side away from a north-west oblique light. Good cliff-drawing. Glaciers. Spot heights in metres, small and rather illegible. Trans-frontier country shown in full. Roads double black lines, white between, showing up well on the dark hachures. Tracks long-dotted, not very distinct. Scale of kilometres and hours. (One hour equals 4·8 km.) Latitudes and longitudes shown on margin.

Sexagesimal and centesimal systems. Origin of longitudes not stated. Meridians not carried across the sheet.

This map deserves study as an excellent example of engraved hachures.

(Sheet XVIII: Rhone Valley and Simplon. Dated 1854, revised to 1907.)

This sheet, described above from its 1907 edition, was very greatly improved in 1910 by taking out the rivers and river names from the engraved black plate, and putting them on a separate blue plate. A pale blue shade was at the same time given to the glaciers, with excellent effect.

This is known as the Dufour Map. No characteristic sheet.

Switzerland. 1/250,000. Special railway map. Published by the Swiss Topographical Bureau, Bern. 1908.

The ordinary engraved and hachured map is printed in brown. Railways with names of stations are heavily overprinted in black. An excellent example of the map for special purposes.

YUGOSLAVIA. 1/75,000. Kingdom of Serbia.

Photolithographed in colours at Vienna for the Serbian General Staff in 90 sheets 14½ by 13¼ inches. Contours in brown at 50 metre intervals with few spot heights. Water black, forests green, roads red and black. All names in Cyrillic.

(Sheet S. 3: Petrovac. No date, but about 1910.)

Yugoslavia. 1/100,000. Vojni Geografski Institut Kraljevine Jugoslavije.

Sheets about 16 by 22 inches covering 30′ by 30′ with longitudes from Paris, and longitude from Greenwich in blue.

Relief by brown contours at 20 metres, each fifth strengthened, with many intermediate contours in pecks and dots, very confusing. Roads and town plans black; water and water names blue; woods very pale green tint. All names in Roman character.

Some early sheets of this new series had red contours at 50 metres.

(Sheet Beograd. 1931.)

Yugoslavia. 1/200,000.

Photolithographed in colours about 1893 in twelve sheets of varying size. Hill-shaded in brown with many spot heights, roads in red and black, rivers in black with Danube filled blue, town plans in black but Belgrade red; forests green. Graticule carried across sheet; longitudes from Paris.

There is a very similar map on 1/250,000 of about the same date but with brown hachures and water in blue.

(Sheet 6: Niš.)

Yugoslavia. 1/200,000.

Serbian edition of the Austrian 1/200,000 with names revised.

Chapter VI

OTHER FOREIGN MAPS

A CHAPTER on Topographical Maps beyond the frontiers of Europe is comparatively short. The maps of the United States are still very incomplete, and there is no general map of the country. Egypt, Japan, and Siam have made good progress. The French, Italians, and Dutch have been active in their overseas possessions. The Russian 40 verst map is in process of revision. A good map of Turkey in Asia is in progress. Apart from these and from the British maps treated elsewhere, there is little to describe.

EGYPT. 1/50,000. Survey Department.

Sheets about 19 by 17 inches, covering 15′ of longitude by 12′ of latitude. No contours, and spot heights rare. Main roads filled brown; green symbol not explained in reference, presumably palms. Limits of desert dotted brown. Elaborate water symbols: river and lakes blue lined; canals intense blue, drains green, all carefully graded. Names in Roman and Arabic.

(Sheet III–I N.W.: Minuf. 1914.)

Similar sheets in the Delta with contours in red at 1 metre.

(Sheet III–II N.E.: Minyet-el-Ramh. 1915.)

Egypt. 1/100,000. Survey of Egypt.

Sheets about 23½ by 16 inches, covering 60 by 40 kilometres. Margin figured in kilometres from origin. Contours brown at 30 metre interval, and rock-drawing. Water blue lined; canals and drains solid blue. Main roads and principal desert tracks red. Names in Roman only, but explanations in Arabic also. A clear and good map rather spoiled by symbols looking heavy owing to roads running along banks of canals.

(Sheet 80/60: Cairo. 1926.)

Sheets of this series in the Delta have contours at 1 metre. Sheets of the Red Sea Coast series have black hachures and lines of drainage green.

Egypt. 1/250,000. Survey Department.

Sheets about 22½ by 17½ inches, covering 90′ by 60′. No contours, and spot heights rare. Water as on 1/50,000. Roads red, rather confused with Markas boundaries orange. Limits of desert dotted brown. A good well-printed map.

(Sheet 1–E: North-West Delta. 1910.)

FRENCH AFRICA. Algérie. 1/50,000. Service Géographique de l'Armée.

Sheets as Tunis 1/50,000, *q.v.*

Algérie. 1/100,000. Service Géographique de l'Armée.

Sheets about 19 by 12 inches. Brown contours at 50 metre interval, and grey oblique hill-shading. Water blue; woods black symbol.

(Sheet No. 29: Thala. About 1913.)

Dahomey. 1/100,000. Service Géographique de l'A.Occ.F., Dakar.

Like Senegal, but without contours. "Le relief est représenté par un estompage purement figuratif."

(Sheet B 31 XV 1: Cotonou. 1925.)

Maroc. 1/500,000. Service Géographique, Rabat.

Sheet 23 by 14 inches, rectangular, apparently on Bonne's projection, graticule figure in grades. Brown contours at 200, 500, and by 500. Grey oblique hill-shading. Water blue, woods lined green.

(Sheet No. 2: Oudjda. 1921.)

Sénégal. 1/100,000. Service Géographique de l'Afrique occidentale.

Sheets 19 by 23 inches covering 50′ by 60′. Brown contours at 10 metres, each fifth strengthened. Water blue; vegetation various green symbols. Metalled roads red; town signs red and black. Distinction between perennial and non-perennial streams, fresh and salt lakes. Scale of kilometres and *milles*, 13 km. equals 7 milles.

(Sheet No. XIII: Fatick.)

Sénégal. 1/200,000. Service Géographique de l'A.Occ.F., Dakar.

Similar in style to the 1/100,000, but with contours at 20 metres. Sheet 1° square with prime meridian Greenwich. In the border are given the distance of the southern edge from the equator, the distance of the south-east corner from the prime meridian, the length of 10′ of top and bottom parallels, and of the meridian, and the area of each 10′ square: very useful information which might with advantage be given on every map. Projection not stated.

(Sheet D 28 XIV: Thiès. 1923.)

Tunis. 1/50,000. Service Géographique de l'Armée.

Rectangular sheets covering 32 by 20 km., considerably oblique to the graticule. Brown contours at 10 metre interval, cliff-drawing and oblique hill-shading in grey. Roads red; water blue; elaborate symbols. Much like the pre-War French 1/50,000.

(Sheet XX. B 2, C 36: Tunis. 1924.)

Tunis. 1/100,000. Service Géographique de l'Armée.

Rectangular sheets about 19 by 12 inches, oblique to graticule. Brown contours at 25 metre interval, and oblique hill-shading. Roads black; water blue; towns red; woods close green symbol.

(Sheet XLVII: Kasserine. Surveyed 1900.)

Tunis. 1/200,000. Service Géographique de l'Armée.

Rectangular sheets about 19 by 12 inches. Brown hill-shading; water blue; detail black. Old reconnaissance series.

(Sheet No. XII: El Djem.)

Afrique. 1/500,000. Service Géographique de l'Armée.

Sheets covering 3° by 2°, on the projection of the International Map, and corresponding sheet lines. Brown contours at 500 metre interval, with oblique hill-shading from north-west. Water and water names blue; green tint for woods.

(Sheet N.J. 32: S.E. Tunis. About 1923.)

Sahara. 1/1,000,000. Service Géographique de l'Armée.

Projection and sheet lines of the International Map. Brown contours at 100 metre interval to 500, 700, 1000, and thence by 500. Water blue; other detail black. Names rather large.

(Sheet N.I. 30: Fes. 1926.)

Other sheets of this series with brown unfigured form lines and sand tinted buff. (Sheet N.F. 31: Fort Laperrine. 1924.)

INDO-CHINA. 1/200,000. Service Géographique de l'Indo-chine.

Sheets about 15 by 10 inches, covering 75 by 50 km.

Contours or form lines brown, unfigured, with rare spot height and rather heavy oblique hill-shading in grey. Roads red; water blue; cultivation green.

(Sheet Bao Lac. 1911.)

A newer edition in large sheets about 30 by 20 inches with contours at 25 metres.

Indo-China. 1/500,000. Service Géographique de l'Indochine.

Sheets about 35 by 20 inches.

Relief by conventional hill-shading in grey and rock-drawing in brown, and spot heights. Roads red; water blue. Provisional in type.

(Sheet Hanoi No. 5. 1923.)

Indo-China. 1/1,000,000.

Conventional hill-shading in grey. Roads red; water blue.

Similar to the old series of Asia, except that the latter are remarkable for their projection, that of the maps of Ptolemy, with straight parallels and meridians inclined inwards. The series covers about half China, parts of Central Asia, Persia, and extends into Europe. Similar sheets of more recent date on 1/2M along the Siberian Railway.

ITALIAN AFRICA. Eritrea. 1/100,000. Carta della Colonia Eritrea.

Sheet 14 by 14 inches. Contours at 100 metres and grey hill-shading. Water blue, tracks red.

(Sheet 11: Asmara. 1909.)

Tripolitania. 1/100,000. Istituto Geografico Militare.

Sheet 18 by 14 inches. Hachures grey, oases green, wells blue, cultivable land buff.

(Sheet 2: Tripoli. 1913.)

A later edition in black contoured at 10 metres.

(Sheet 13: Bengazi. 1928–Anno VI.)

Tripolitania. 1/200,000. Ministero delle Colonie.

Sheet 21 by 16 inches. Brown hachures, blue water, red tracks.

(Sheet Zuara. 1920.)

Tripolitania. 1/400,000. Schizzo dimostrativo della Tripolitania Settentrionale.

Sheet 20 by 12 inches. Brown form lines, water blue, tracks red.

(Sheet Zuara. 1916.)

Tripolitania. 1/800,000. Ministero delle Colonie.

Sheet 28 by 22 inches. Brown hill-shading, water blue, tracks red. Stony desert buff, sandy yellow.

(Sheet 1: Tripoli–Gadames. 1927.)

Somalia. 1/50,000.

Sheet 22 by 14 inches. Contours at 10 metres. Vegetation yellow and black symbols, desert buff.

(Sheet Torda. 1911.)

Somalia. 1/200,000. Istituto Geografico Militare.

Sheet 22 by 15 inches. Brown form lines, water blue.

(Sheet Mogadiscio. 1910.)

Somalia. 1/400,000. For R. Governo della Somalia Italiana by Agostini, Milano.

Sheet 22 by 22 inches. Yellow ground tint, water blue, tracks red. List of astronomical positions.

(Sheet 16: Mogadiscio. No date, but recent.)

Cirenaica. 1/50,000. Istituto Geografico Militare.

Sheet 18 by 14 inches. Black. Contours at 10 metres.

(Sheet Derna. 1920.)

Cirenaica. 1/400,000. Serv. Cart. Ministero delle Colonie. Carta dimostrativo della Cirenaica.

Sheet 31 by 20 inches. Few black hachures, buff ground tint. Mercator's projection.

(Sheet 1: Altopiano Cirenaica. 1928–Anno VI.)

EAST CHINA. 1/1,000,000. Kart. Abt. K. Preuss. Landesaufnahme.

Sheets covering 6° by 4° on the same Ptolemaic projection. Hill-drawing brown; roads red; water blue. Names in Chinese and German transliteration of Chinese. Complete in 22 sheets.

(Sheet No. 7: Peking. 1909.)

A similar map of Chili and Shantung on 1/200,000 with names in Chinese and German transliteration. (Sheet G. 10: Yung Ping fu. 1907.)

JAPAN. 1/50,000.

Sheets about 19 by 14½ inches: 15′ by 10′.

All in black and in Japanese only. Contours at 20 metres.

A more recent sheet of Korea on the same scale has brown contours and cliff-drawing; roads red; woods green; water blue. Names in Japanese and Korean transcribed into English.

(Sheet Keum-Gang-San–Diamond Mountain. About 1919.)

Japan. 1/200,000. Imperial Geological Survey.

Sheets about 18 by 11 inches, covering 60′ by 30′. Contours grey at 40 metre intervals. Roads black; water blue. Names transcribed and elaborate characteristic sheet in English.

(Sheet Osaka. 1920.)

Another edition with hill-shading green over grey contours; towns red. Names in Japanese only. (Sheet Yokosuka: Land Survey Department. 1927.)

Japan. 1/500,000. Land Survey Department.

Sheets about 18½ by 14½ inches covering 60′ by 40′. Hills in greenish hachures; roads red; water blue. Names in Japanese.

(Sheet Tokushima. 1924.)

NETHERLANDS INDIA. Java. 1/100,000. Topografische Inrichting.

Sheets about 14½ inches square covering 20′ by 20′ but without graticule.

Relief by brown contours at 50 metre interval, each tenth strengthened, and spot heights; oblique hill-shading. Roads red and yellow; water blue; cultivation green. Elaborate characteristic sheet. Mapped in Residencies, and stopping short at the boundaries.

A nice clean map rather spoiled by too prominent "Pasir" names. About 25 sheets published.

(Sheet 35. No name. 1924.)

Java and Madoera. 1/250,000. Topographische Inrichting, Batavia.

A very similar map with form lines at 100 metre intervals to 600 and 200 above, and layer coloured.

Java and Madoera. 1/500,000. Topographische Inrichting, Batavia.

In four sheets about 21 by 31 inches.

Relief by form lines at 100, 250, and thence by 250 with layer colours. Roads red; water blue.

A clear but rather coarsely drawn map. Replaces an earlier edition with hill-shading but no form lines.

The above are typical of the Dutch East Indies sheets, which are in many series, of varying completeness.

RUSSIA-IN-ASIA. 1/168,000. Russian General Staff.

Sheets 23 by 20 inches covering all Russian Asia. Conventional brown hill-drawing. Roads black; water blue; sand red. As in all Russian maps the smaller names in Russian script difficult to read. Graticule on a conical projection very much inclined to sheet margins when far from central meridian.

(Sheet XIV: Urga. 1922. Moscow Cartographical Section, Corps of Military Topographers.)

SIAM. 1/50,000. Royal Survey Dept. of the Army.

Sheets about 14 by 14 inches, covering 10′ by 10′. Contours brown; roads red; water blue; woods green tree symbol. Names in Siamese.

Siam. 1/100,000.

Sheets same size covering 20′ by 20′. Conventional brown hill-drawing; water blue; roads red. Names in Siamese.

TURKEY-IN-ASIA. 1/200,000. Turkish General Staff.

Rectangular sheets about 20 by 16 inches, with graticule on a conical projection with central meridian Istanbul; hence graticule much inclined to margins in the east of the series. Brown contours at 50 metres, each fifth heavily strengthened. Roads black; water blue; woods green tint. Names in Turkish. 10 km. grid in red.

(Sheet Ispir. About 1918.)

An earlier edition with conventional but well executed hill-shading in grey instead of contours. A good reproduction in Russian of all the eastern sheets, on 1/210,000 (5 verst).

UNITED STATES. 1/62,500. Geological Survey.

Sheet covers 15′ of longitude and latitude. Brown contours at intervals of 10 feet, each fifth strengthened. Spot heights in black and in brown. Roads black, water blue; woods green tint.

Well drawn clean maps, effective owing to small number of names.

Margins figured to 5′, and sheets divided into townships and ranges of six miles a side. Explanation and conventional signs on back.

(Sheet St Francis, Minnesota. 1919.)

Sheets on this scale in scattered blocks, with contour intervals varying from 5 to 100 feet.

United States. 1/125,000. Geological Survey.

Sheet covers 30′ by 30′. Brown contours at 100 feet interval. Similar in style to above, and for scattered areas not covered by the larger scale.

(Sheet Corona, California. 1902.)

United States. 1/250,000. Geological Survey.

Sheet covers 60′ by 60′. Brown contours at 100 feet interval, every fifth strengthened. Similar in style to above, and for areas not covered by either of the larger scales. Contours very effective in absence of other detail.

(Sheet Ballarat, California–Nevada. 1913.)

About one-half the area of the United States is published on one or other of the above scales, and certain areas on 1/24,000 or 1/31,680. Note that in the catalogues the scales are described inaccurately as 1 inch to 1 mile, instead of 1/62,500.

Chapter VII

CHARTS AND ATLASES

THE style of map-production does move with the times, though slowly; the style of charts does not move at all. They are still all in black, except for a little hand-colour to mark lights, and a pink wash to show where revisions have been made in the edition; they are crowded with soundings expressed in very illegible numerals, and show little evidence that anyone has given a thought to their appearance. They are individuals, never forming sheets of a series, and on very odd scales. They confine themselves strictly to navigation and its dangers, show nothing of the land except what will serve as a mark at sea, reckon their soundings from low water at spring tides, confine their contours to shallow water, and except for the tracks of exploring ships on polar charts pay little or no attention to history. They are printed on very good paper which will stand the use of pencil and rubber and parallel rulers, and are no doubt perfectly adapted to their practical purpose, except that better design of lettering would make them much more legible. With the exception of the polar charts, and of a few gnomonic for projecting great circle courses, they are nearly all on Mercator's projection.

Until quite recently atlas makers have been very conservative also, though they have freely used lithography in colour. The supposed necessity for showing as many names as possible has led to overcrowding, and the retention of the illegible character stump for all the smaller names, with the sad colour belonging equally to the middle of the last century, makes most of them depressing. The heavy smooth paper suited to lithographic printing and strong enough to stand continual handling in large sheets is unpleasant in quality, and what is oddly called a Hand-Atlas may weigh up to sixteen pounds.

But it must be remembered that atlases, unlike the maps with which we have had to deal, are commercial products, and unless they are subsidised, must be made to pay. They are excessively expensive to produce, and if they are to be up to date require a large staff of compilers always at work collecting new detail from geographical publications. And recent events leading to a strong nationalist desire to change accepted names, especially by reversion to original but unfamiliar forms, have required immense changes in the maps of Europe.

The following notes mention what may fairly be called the best atlases in the market. It is unfortunate that in 1933 there is to be bought no atlas of quite the first class in the English language. The large atlas of Bartholomew which was put on sale by a newspaper and afterwards came into the hands of a department store is no longer obtainable. The atlases of Andree and Stieler are well compiled and revised frequently but far from clear. The fine atlas of the Touring Club Italiano unhappily transliterates familiar names like Shyok into Sciaioc and so is less useful than it might be. The most original modern atlas, the Swedish *Nordisk Världsatlas*, splendidly produced, is out of print and will not, it is said, be reprinted. It should be sought for collections for its fine printing and its interesting projections.

"The Times" Survey Atlas of the World. J. G. Bartholomew, 1920.

Many maps layer coloured in green and brown; contours black, dotted; rivers black. Political maps tinted, with ribbon boundaries. Full separate index. Out of print. Weight 18 lbs.

Philip's International Atlas, 1931.

Most maps tinted politically; a few layer coloured. Relief by black hachures. Rivers black. Large index included. Weight 12 lbs.

Nordisk Världsatlas, by S. Zetterstrand and Karl D. P. Rosen. Nordisk Världsatlas Förlag, 1926.

A few maps simply layer coloured, but most with grey hachures on buff ground-tint and boundary ribbons in clear bright colours. Rivers blue but names black; sea layered blue with blue names and soundings. Ice bright blue. Railways black. Interesting projections carefully named and described. Illustrated descriptive chapters and full index. Weight 9 lbs.

Atlante Internazionale del Touring Club Italiano, by L. V. Bertarelli, O. Marinelli, and P. Corbellini. Milano, 1929–VII.

A few maps layer coloured; rest with brown hachures. Ice greenish blue, railways red, rivers black. Well printed on matt paper, probably by offset. Weight $17\frac{1}{2}$ lbs.

Atlante Universale. Istituto Italiano d'Arti Grafiche, Bergamo, 1927.

Purplish brown hill-shading. Rivers blue; sea layered. Extensive text with plans. Projections named. Bright attractive colours. Weight 8½ lbs.

Andrees Allgemeiner Handatlas, by Dr Ernst Ambrosius. Velhagen und Klasing, Bielefeld und Leipzig, 8th Edition, 1928.

A few layered maps, but mostly with brown hachures and colour-ribbon boundaries. Rivers black. Names Germanised (real in brackets). Some projections named. Extensive separate Index, one of the best. Weight 14 lbs.

Stielers Hand-Atlas, by Prof. Dr H. Haack. Justus Perthes Geogr. Anstalt, Gotha, 10th Edition, 1930.

Relief by purple hachures; coloured ribbon boundaries. Rivers blue, sea greenish blue layered, deserts buff. Projections not named. Extensive index included. Weight 12 lbs.

Atlas Universal de Géographie, by F. Schrader. Librairie Hachette, 1923.

Purple hachures and coloured ribbon boundaries, with a few maps layer coloured. Rivers black, sea green. Projections not named. Names in French form: Edimbourg, Varsovie. Weight 16 lbs.

Atlas Colonial Français, by Commandant P. Pollacchi. *L'Illustration*, 1929.

Clear maps with brown hill-shading; rivers blue. Text, and many photogravure illustrations.

It is greatly to be desired that someone should take in hand an entirely new atlas, lettered in something like the new Ordnance Survey alphabets, with borders and titles designed to be quietly decorative, printed by the offset process on antique paper, with light clean layer colours and hill-shading, and names as in actual use in the country concerned, with the conventional names admitted for British official use by the P.C.G.N. in brackets. There is much room for enterprise in choosing interesting projections.

Chapter VIII

EXPLORATORY SURVEY

The explorer's route map.

THE first care of a traveller who passes through an unknown, or but partially explored, country is to make a record of where he has been, and of the main features of the country along the route by which he has travelled. Often single-handed, encumbered by transport, compelled to keep to the track, and unable to leave his party, he cannot hope to make anything in the nature of a map, in the ordinary sense of the term. But for his own guidance, to avoid getting lost, he is compelled to determine his position day by day in much the same way that the position of a ship is determined at sea, by observation of the Sun and the stars, so that he is able to say roughly in what latitude, and perhaps in what longitude his halting places were. Moreover, as he goes along he is able to make such observations of the shape and course of his path as to enable another man coming after him not only to arrive more or less at the same place, but to follow the same route. And finally, he can keep a sort of running record of the things that lie on each side of his path. All this he does by the construction of a "route traverse" or "route map".

It is essentially the work of the pioneer, whose main business is to get through the country, and who can afford to give to mapping and survey only a small part of his attention, and no voice in the determination of his plans. Such is the first exploratory survey of a country.

The necessity for this kind of survey is of course rapidly diminishing with the progress of triangulation. A modern traveller will take care to provide himself with the figures for all the fixed points in the country, and will be less dependent on his own unaided efforts. He will often be able to control his route traverses by making them start and close on fixed points. But there are still vast territories in which his controls must be his own astronomical fixes.

The astronomical observations.

These differ very little in kind from those used in much more elaborate work, and we may defer a detailed consideration of them, confining ourselves for the moment to an examination of the general principles, which are common to voyages on sea and on land.

Latitude, the distance north or south of the equator, is most easily found by the observation of the altitude of a heavenly body as it crosses the meridian: the Sun about noon, or stars at their meridian passages during the night.

Longitude, the angle between the meridian of the observer and the standard meridian, generally that of Greenwich, is measured by the difference between his local time, determined by his own observations, and the time of the standard Meridian carried by his chronometers; and unless the traveller is travelling very light, he will usually be able to control his chronometer frequently by wireless time signals, being in that respect vastly better off than his predecessors.

Azimuth, or true bearing, required for the correction of the compass, and for resection of position from fixed points, is conveniently determined with the local time.

At sea the old-fashioned mariner works almost entirely by the Sun, and the observation which is familiar to all passengers is the noon altitude of the Sun, which gives the latitude. It is a very common mistake to suppose that this observation gives the instant of noon. This is not so; the observation which gives local time is made somewhat early in the morning or late in the afternoon. Thus latitude and longitude are determined at different times. But meanwhile the ship is under way, and it is necessary to have some method of carrying forward the morning longitude to the latitude at noon, or of carrying forward the noon latitude to the afternoon longitude. This is done by keeping the "dead reckoning" of the ship's course. A continuous record of the course and of the speed of the vessel on that course is kept in the log (the journal of the voyage, not the instrument of the same name by which the speed is measured). The dead reckoning enables the navigator to make the required allowance for the run of the ship between the successive observations, and to carry on during cloudy weather.

The traveller on land works in very much the same way. There are obvious reasons why it may be necessary to avoid as far as possible observation of the stars by night: mosquitoes and the risk of fever are sufficient. Though star observation is more accurate and is increasingly used, he may have to rely largely on observations of the Sun. But the requisite observations of the Sun must be made, as we have seen, at widely different times of day, and the traveller cannot as a rule afford to wait for a great part of a day at every place whose position he wishes to fix. He is therefore compelled by the nature of the case to use some means of keeping account of what the sailor calls his "dead reckoning", that is to say, of keeping a current account of his position, carried on from one place to another by observation of the course he is steering and the rate at which he is travelling. He makes, in fact, what is called on land a "compass traverse".

In modern practice the navigator does not confine himself to the old routine of the noon altitude for latitude and the morning or evening sight for time and longitude, but in doubtful weather he gets an observation whenever he can. An altitude of sun or star, at a known standard time by chronometer, fixes a small circle of the sphere on which the observer must be at the moment of the observation; and a portion of the circle sufficient to cover the range of possible positions can be laid down as sensibly a straight line on the chart. A subsequent observation made on a different bearing defines another small circle, which intersects the first at the position of the observer. Intelligently used this method is of wide generality, and has the great merit of giving its full weight to any observation made at any time, while avoiding the often troublesome necessity of stopping to make the observations at closely defined instants. It is now employed with great advantage upon land. Within a few degrees of the poles its application becomes of remarkable simplicity. See, for example, papers by the author in the *Geographical Journal* for March 1910 and July 1926.

The compass traverse.

A traverse is defined as a connected series of straight lines on the Earth's surface, of which the lengths and the bearings are determined. In a compass traverse the bearings are determined with the prismatic compass, which differs from an ordinary pocket compass in two principal respects: it is fitted with sights which can be directed upon a distant object; and with a prism which brings into view at the same time the scale of degrees marked round the edge of the compass card. The bearings are invariably reckoned in degrees from 0° right round to 360°, from magnetic north through east. The complicated system of "points", now becoming obsolete at sea, is never to be used on land.

Selecting the most distant conspicuously recognisable point upon the line of march, the traveller observes and records its bearing, and proceeds to determine, as accurately as circumstances will allow, the distance he has travelled by the time he arrives at it, or the length of the "leg". Accurate measurement is of course inconsistent with rapid travel. A short distance can be paced, but it is not possible to count paces all day for days at a time. The legs of a traverse are therefore generally measured by cyclometer or "perambulator", or merely estimated by time.

Distances by cyclometer or perambulator.

The perambulator consists of a wheel of known circumference, frequently ten feet, mounted in a fork with a handle, very much like the common child's toy, and fitted with a counting mechanism to record the number of turns which the wheel has made. If a man can be spared from the caravan to trundle this instrument—an easy duty which fills the native carrier with pride—it is simple to record the length of each leg of the traverse in terms of the number of revolutions; but it is hard to tell how much to deduct for the windings of the path and the inequalities of the ground.

The sledge meter used by Sir Ernest Shackleton and Captain Amundsen on their South Polar journeys was a "perambulator" wheel carried out on a light spar behind the sledge. Its indications were in general remarkably accurate.

The ordinary cyclometer registers only eighths or tenths of a mile, and is too coarse; a more refined instrument of the kind would sometimes be useful. But it is easy to calibrate any bicycle so as to know the value of a revolution of the front wheel.

An interesting recent development of this kind of traverse is found in the methods of the Desert Surveys of Egypt, originating in the motor reconnaissances during the war, and elaborated by Dr John Ball.

Distances by time.

The apparent advantages of the perambulator method of measuring distances are in practice much discounted by the difficulty of knowing what allowances to make for the windings of the path. An experienced traveller will obtain results which are pretty well as good, without being obliged to spare a man for the work, by the simple process of timing each leg of the traverse, and estimating the rate of march. The average rate of a party on fairly level and unobstructed ground is found to vary very little. Practice will enable the traveller to make allowance for change of rate over rough or hilly country. The most common error is a persistent over-estimation or under-estimation of the rate; and we shall see later how it is possible to keep a check on systematic errors of this kind. The most serious difficulty is common to all methods of route traverse—that of keeping the average direction and estimating how fast one is really covering the ground, when marching along a winding track through thick bush.

Details on the flanks.

If the country through which the traveller is marching is fairly clear he can fix roughly the positions of its principal features as he goes along by the method of cross bearings. The compass bearings of prominent peaks and other easily recognisable objects are taken from different points, and the intersections of these lines of bearing, when plotted at the end of the march, give positions of the objects in question with a degree of accuracy comparable with the general accuracy of the traverse.

At the same time the bearings of all cross tracks and streams are taken at the points where they meet the line of march; the character of the ground is recorded at intervals, with any other information which can be obtained readily.

Check by cross bearings.

A similar process can be used for the opposite purpose, of checking the accuracy of the traverse by bearings of a distant object whose position is known. For example, in traversing round about Mt Elgon occasional bearings of the peak will serve as a check on the traverse, and lay down its position on the ground. A like method is much used at sea for fixing the position of the ship when a known object is in sight, such as the Peak of Teneriffe, visible sometimes at a distance of 100 miles.

The field book.

The method of recording a traverse is best shown by a page of the field book, which is kept by well-recognised rules.

The essential is that it begins at the bottom of what would ordinarily be the last page of the book, and goes upwards and backwards to the beginning. The columns up the centre contain the time of each observation, the bearing of each leg, and the estimated rate of progress on that leg, or the number of turns made by the perambulator wheel. Detail to the right or left flank is recorded in the right or left margin, and it is important to note that the page becomes a kind of diagrammatic representation of the country traversed, of which the central columns represent a line, the line of march. Hence if a straight stream or track crosses the route obliquely, the portions represented on each flank must not be drawn as parts of one straight line, but in the manner shown in the example below. A similar convention will be found later in the methods of keeping field books for other kinds of survey.

Village W

	Time	Compass Bearing	Rate M.P.H.	
Peak A 127°	11.14			
			3	
	10.52	198		
			$2\frac{3}{4}$	
	10.41			
	10.34	220		Shallow stream 10 ft. wide
			3	
Halt at Z. Shade trees and good water	10.7	234		
	8.52			
			3	
	8.40	202		
			$2\frac{1}{2}$	
	8.22	219		Track to Y 308°
			$2\frac{3}{4}$	
	8.5	223		
			$2\frac{3}{4}$	
	7.57	246		
			3	
Snow peak A 146°	7.46	214		Cultivation

Leave Camp at Village X

Specimen field book.

The compass.

For instrumental details as to the care and use of the compass, see page 142. We shall deal here with the particular points which are special to compass traversing.

It is usually essential that the march of the party shall not be interrupted while the observations are made, and it is generally undesirable that the observer should have to run in order to catch up with the party after the observation. If he is on foot he will try

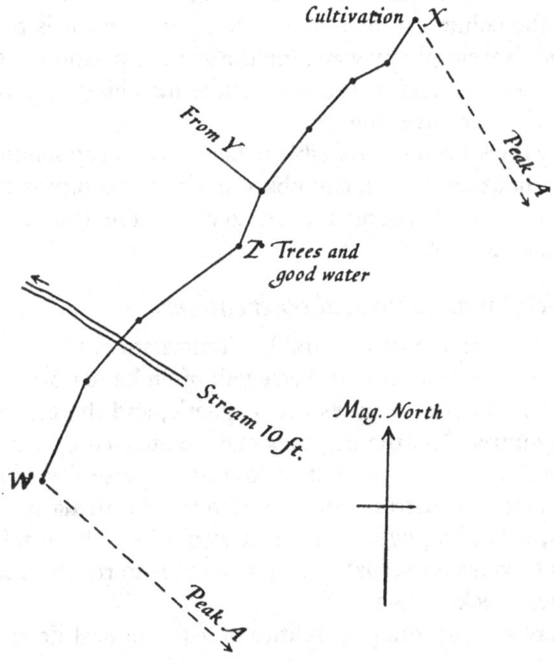

Fig. 2. *Plot of compass traverse.*

to walk on ahead to the point of observation; if he is mounted he can afford to spend more time, perhaps to dismount and set up the compass on a tripod, which much improves the accuracy of the observation. It is clearly impossible to prescribe any exact rule or programme. Two points are to be remembered: the rate of march which is recorded is that of the main party, which keeps on steadily, and it is therefore unnecessary for the observer to take account in his time records of the short intervals during which he himself is halted to make the observations. And secondly, the whole operation is a rough and ready affair; he must therefore be careful not to waste time in trying to record minute details which have no real importance; small deviations of the track will be ignored so long as the general direction is preserved, and it is not necessary to take a careful bearing of a cross track which disappears round a corner in a few yards.

In thick bush, where the direction of the path changes every few yards, or on the march in a hostile country, when it is impossible to leave the column of march even for a moment, it is possible to do a good deal simply by watching the average position of the compass card as it is held in the hand while marching, and recording the bearing every five minutes.

In very thick bush it is possible to take bearings by sending a man ahead to shout or whistle, and observe with the compass the direction from which the sound appears to come. The results are much better than one would expect.

The check by astronomical observations.

Even under the most favourable circumstances the error in the recorded length of a route traverse will often be ten per cent.; and in a country where the rocks are magnetic, and the compass consequently unreliable, the bearings may be affected by large errors. It is therefore very important to lose no opportunity of checking the traverse by astronomical methods. Particularly on long journeys, extending over months, it would be folly to rely on the compass traverse only, just as it is dangerous to rely for many days on the dead reckoning at sea.

We may sum up the possibilities of astronomical determination as follows:

Latitude. Observation of latitude, either by the Sun or the stars, is easy, and there is no difficulty at all in getting latitude within a mile by sextant, or within a small fraction of a mile by micrometer theodolite.

Longitude. Observations to find local time are easy, though the calculations are a little long. But the longitude is the difference between local time and Greenwich time, and the former practical impossibility of getting longitudes right within a number of miles while upon the march was due to the difficulty of carrying or obtaining Greenwich time. To carry Greenwich time meant to carry such a number of chronometers, or better, of half-chronometer watches, that the mean of their indications, corrected for their rates so far as known, is right within a small number of seconds. A difference of four seconds of time is equivalent at the equator to a difference of one geographical mile.

The cost of the chronometers, the anxiety of their care and transport, and the little reliance that could be placed upon their rates when they were exposed to jolting and great changes of temperature, made their employment practically impossible. Watches carried carefully in the pocket are a little more satisfactory, but cannot be trusted absolutely for long.

The alternative was to find Greenwich time by the astronomical observation of occultations of stars or of lunar distances. Both methods involved skilled observation and long calculations, and they are now happily obsolete.

Since the Greenwich time signal from the Rugby wireless station was established in the autumn of 1927 there is no part of the world in which a traveller equipped with relatively modest apparatus cannot receive Greenwich time twice a day with relative ease from one or other of the principal stations: Rugby, Bordeaux, Nauen, Annapolis. He will then have to trust his watch only to keep uniform time between the signals.

Azimuth.

The azimuth or true bearing of a ray, the angle between the ray and the meridian through the line of sight, is different from the compass bearing of the ray by the amount of the deviation of the compass. Were this deviation constant it would have the effect of slewing all compass work round in azimuth, but not of altering its shape. But the deviation of the compass varies not only from place to place, but also to some extent from month to month at the same place. The compass has also a diurnal variation of nearly half a degree, more or less according to the phase of the sunspot cycle, and occasionally much more during magnetic storms. Moreover, the compass is very likely to be disturbed by magnetic rocks. Hence the compass is at best an uncertain instrument that cannot be relied upon within half a degree; and no compass work is complete unless it is accompanied by determinations of the deviation at the place and time. Such observations consist in finding from the Sun or the stars the true azimuth of a given ray, and comparing it with the compass azimuth.

The determination of true azimuth is made in much the same way, and in the same circumstances, as the observation for local time, and if necessary can be combined with it. It is not a very laborious process, and should therefore be practised frequently.

At sea the observation for the deviation of the compass is more frequent than any other observation; on a well-run ship an observation is taken every watch, if the weather allows, for an unknown error of even a quarter of a degree is by no means negligible in the day's course of a fast steamship.

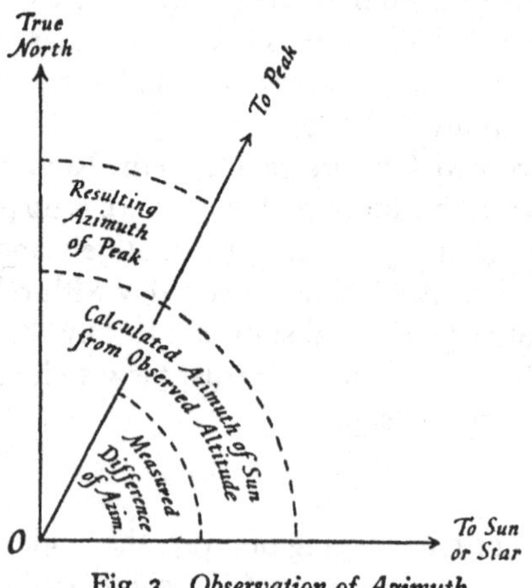

Fig. 3. *Observation of Azimuth.*

On land, where the rate of progress is much slower, and the compass used is smaller and less accurate, such very frequent control is not necessary. But it is essential to control the general accuracy of a compass traverse by taking a true azimuth from time to time, and an example will illustrate the whole process.

Before starting in the morning the traveller may be informed by his guides that the day's march will take him near a distant well-marked point. He sets up his theodolite, observes on the distant point, and reads the horizontal circle. Then he turns to the Sun and obtains simultaneous readings of the horizontal circle and the Sun's altitude. From the altitude, and an approximate knowledge of the latitude, the true azimuth of the Sun can be calculated. Apply to this the angle between the Sun and the distant point, as measured on the horizontal circle, and the azimuth of the distant point is found. Compare this with the compass bearing of the point, and the deviation of the compass is known.

Now when at the end of the day's march the route is plotted, and the position of the object observed to in the morning has been laid down with reference to the route by cross bearings taken from near at hand, the azimuth of this object from the starting point, as shown on the drawing, may be compared with its true azimuth as found from the morning observation. This will provide an excellent check on the general accuracy in azimuth of the whole day's traverse. It must not be forgotten that, except near the equator, the "convergence of the meridians" must be taken into account in plotting a long traverse, but the discussion of this refinement may be postponed.

Check on the length of the traverse.

We have seen that latitudes may be found very easily to within a fraction of a minute of arc, that is, of a geographical mile. A comparison between the latitudes taken at each end, and the distance made north or south, as shown by the traverse, will give an admirable check on the scale of the traverse in this direction.

But it should be noted carefully that before proceeding to resolve the traverse into its north-south or east-west components, all the bearings must be reduced from magnetic to true bearing. The resolved parts of each leg of the traverse may then be calculated directly, or they may be taken out from Traverse Tables, or from the tables which are called, in the curious terminology of the navigator, "Latitude and Departure Tables".

A similar check on east and west traverses can of course be obtained from astronomical longitudes if wireless time signals can be observed. If not, the check by differences of latitude whenever the route leads considerably north and south will give confidence in the length of the traverses running east and west.

Dead reckoning.

The observation for time or azimuth, morning or evening, requires a knowledge of the latitude, which is found from the Sun only near noon; while the observation for latitude requires a knowledge of the local time, except in rough sextant work, and of the approximate Greenwich time, which cannot be found near noon by any practicable field methods. Suppose the traveller has obtained an observation for time before setting out in the morning. By noon he will probably have moved into a different longitude, and his local time will have changed. But his route traverse will give him very nearly the change of longitude, and so he can apply the necessary correction to the error of his watch which he found in the morning, and thus he can obtain his true local time near noon. In the same way, he can bring forward the noon latitude to give him the approximate latitude required for the time or azimuth observation morning or evening.

It will be seen that this process of carrying forward the change of latitude or longitude between one observation and the next is very like the process practised continually at sea, of carrying on by dead reckoning.

The details of these astronomical processes need not be studied at the present stage. But it seems to be essential that the student should have a general idea of the nature of the control which field astronomy can exercise over traverses plotted by the compass, such as the pioneer traveller makes in his first journey through an unmapped country.

Route traverses cannot make a map.

A route traverse carefully made is admirably adapted to illustrate the account of a journey, and to enable future travellers to follow the same route. But it is altogether wrong to suppose that anything in the nature of a satisfactory map can be made by combining a number of these traverses. Each individual traverse will serve very well by itself, until it comes to be fitted to other traverses. Then it is invariably found that the separate traverses will not fit accurately together. The reason for this is easily seen. Each traverse is a zigzag line, which may have been stretched by error in estimating distances, and distorted by errors in the mean bearings of tortuous tracks. A certain number of points will have been tied down, so far as displacements north and south are concerned, by the astronomical latitudes, and the general bearings of considerable lengths will have been controlled by the astronomical azimuths. But within these constraints there will always have been plenty of room for errors to accumulate.

Moreover, a number of traverses run across a country leave large areas unvisited, and a map cannot be considered worthy of the name which shows detail in one part, and leaves out more important detail of the same kind in another. Thus the compilation of traverses cannot make a map.

The impossibility of making a map by compiling traverses is one example of a general principle which underlies the whole of survey: a map must be constructed on a rigid framework. And how can such a framework be obtained? The answer is the same in all branches of surveying: By building it up of triangles in which the angles are measured, not the lengths of the sides.

It must be understood, however, that when we speak of the impossibility of making a map by the compilation of route traverses, we mean that a final and accurate map cannot be made by such means. A study of the maps of Africa issued by the Geographical Section of the General Staff will show that many of these are compiled from such material; but such sheets always bear the legend "None of this country has been surveyed"; the compilation is provisional, a little better than nothing at all, and as soon as possible it is superseded by a regular survey.

We must make the reservation also that there are cases in which the final work of survey must be done by traversing: in dense forest regions, and in cities. But this is precise traverse, of an altogether different order of accuracy, which will be dealt with briefly in Chapter IX, p. 200.

Heights by aneroid barometer.

The clinometer is good for determining relative differences of height over a small range of ground, but is useless for carrying forward such determinations over a long distance: the accumulation of error would be soon intolerable. The traveller therefore requires some instrument to give him rough determinations of absolute height at a point, and to measure considerable differences of elevation under such circumstances as the ascent of a mountain peak, where the clinometer is quite inapplicable.

The pressure of the atmosphere varies with the height above sea, and the reading of the barometer varies accordingly. Roughly, the barometer falls one inch for each thousand feet of ascent. But the pressure of the air and the height of the barometer are affected also by the disturbances moving in the atmosphere which affect the character of the weather; and they are also dependent upon the temperature of the air. Hence the reading of the barometer at any moment is dependent upon a complication of circumstances, and can give no precise determination of height above sea. One may say, however, that the barometer at sea level is rarely above 31 inches, and rarely below 29; so that if it is observed to stand at 24 inches, the presumption is that the observer is somewhere between 5000 and 7000 feet above sea.

To obtain a clear understanding of the way in which the barometer can be used to obtain more precision than this, consider the case of a recording barometer carried by train from Lancaster to Carlisle. At Lancaster it will draw a trace showing the variations in the pressure of the air associated with the passage of disturbances or the establishment of anti-cyclones. Between Lancaster and Carlisle the train climbs nearly 1000 feet over Shap Fell; the barometer will fall about an inch on the ascent, and will rise again as the train runs down the steep descent through Penrith. At rest at Carlisle the barometer will draw another variable trace, depending again upon the passage of disturbances in the atmosphere. The question is therefore, how to disentangle the variations due to height from those due to weather, including changes of temperature.

It is not safe to assume that the sea-level pressure is the same at two places fifty miles apart. Therefore if one wishes to disentangle the weather changes from the altitude changes, it is almost necessary to have a barometer stationary in altitude, as nearly as possible below the barometer which is being carried uphill. The weather changes of pressure are given by the former and applied to the latter; what is left of change may be ascribed to the variation in height of the travelling barometer.

In many cases it is not possible to leave a barometer at the base camp, to be read while the travelling barometer is away. One must then do the best possible by returning to the base camp as soon as possible after the ascent, and determining the change in the reading there which has taken place during the day. Thus, suppose that a climber reads his barometer at 4 a.m. before setting out, and records frequent readings during the ascent and descent. He reaches the top at noon, and is back in camp at 5 p.m. Comparison of the morning and evening records at the camp shows that the barometer has fallen half an inch during the time the expedition was away, and to compare with the noon reading at the summit one must interpolate between the morning and the evening readings below. If the barometer has been falling regularly throughout the day, this will give the correct noon reading at the lower station; but otherwise not; and the advantages of having a second observer remaining below are sufficiently clear.

Barometer heights in exploratory survey.

We have seen that accurate results can be obtained only when the barometer is used to obtain differences of height between two stations which are occupied as nearly as possible simultaneously. In general this is not practicable for an explorer, who has to push on through a country, and cannot retrace his steps, or leave another observer behind at a base camp. Such a man must do the best he can to obtain information as to the average pressure at sea level in the region where he is travelling, and must be careful to make allowances for the seasonal and daily variations of the pressure, which in tropical countries are often surprisingly regular, and quite worth taking into account.

In the survey of Southern Nigeria, for example, it is found that the diurnal changes of the barometer are so regular that it is possible to run fairly accurate contours in the thick forest, if care is taken to study and apply the corrections for the diurnal variation.

But with all care it is not possible for a traveller single-handed to obtain much accuracy with the aneroid barometer, and this explains why the heights of mountains and lakes in Africa are often found to be several hundred feet wrong, when the early barometer heights are at last compared with the results of precise surveys.

Instrumental precautions.

The aneroid is a delicate instrument, and must be treated with all care. It must be allowed some minutes to come to rest before a reading is taken after a rapid change of altitude; the process is hastened by gently drumming with the fingers on the case of the instrument; but hard tapping is bad for the instrument, and all shocks must be avoided.

An ordinary aneroid carried for a long time at a great height, say over 9000 feet, is liable to become strained, and its readings inaccurate. Special types of instrument, called Mountain Aneroids, are made for use at great heights, but it is doubtful if they are really more reliable than the modern make of ordinary aneroid, which has been much improved of late years.

Temperature corrections.

The aneroid barometer is corrected for the effects of temperature upon the instrument itself; but it is necessary to take into account the temperature of the intervening air between the upper and the lower stations. This cannot be done accurately, and the method suffers in consequence. It is seldom possible to do more than to take the mean of the temperatures at the upper and the lower stations, and to treat this as the mean temperature of the intervening column of air. When there is no lower station for comparison, and some assumption has to be made as to the temperature below, the results become still more uncertain.

Heights by barometer.

There are two sets of tables in common use, those calculated respectively by Loomis and by Baily. Both are given in *Auxiliary Tables of the Survey of India*; the former are given in *Hints to Travellers*, published by the Royal Geographical Society; and the latter in *Textbook of Topographical Surveying* (Close).

In none of these places is there an adequate account of the basis of construction of the tables, and it does not appear that any such account is readily available. A brief summary is therefore given here.

An investigation of the theory of the subject is to be found in Laplace, *Mécanique Céleste*, vol. IV, p. 289 of the edition of 1805. With some modifications, this leads to the following formula:

If H = height of barometer at lower station,
H' = ,, ,, ,, ,, upper ,,
T = temperature of *barometer* at lower station,
T' = ,, ,, ,, ,, upper ,,
t = ,, ,, air at lower station,
t' = ,, ,, ,, ,, upper ,,
λ = the latitude,
s = height of lower station above sea level,
x = difference of height of two stations,
μ = modulus of common system of logarithms,
θ = difference of expansion between mercury and the metal of which the barometer scale is made,
a = radius of the Earth.

Then $$x = P \times \{\log H - \log H' - \mu\theta\,(T - T')\}$$
$$\times \left\{1 + \frac{t + t' - 64}{900}\right\}$$ [when temperatures are on the Fahrenheit scale]
$$\times \{1 + 0{\cdot}00265 \cos 2\lambda\}$$
$$\times \left\{1 + \frac{2\mu P + x + 2s}{a}\right\}.$$

Taking heights in feet, the constant P is 60159, according to Loomis; $a = 20{\cdot}89 \times 10^6$ feet, and the constant multiplied by $\mu\theta$ is 2·341.

Hence

$$x = \{60159\,(\log H - \log H') - 2{\cdot}341\,(T - T')\}$$

$\times \left(1 + \frac{t + t' - 64}{900}\right)$ to allow for the mean temperature of the air and an average amount of aqueous vapour

$\times (1 + 0{\cdot}00265 \cos 2\lambda)$ to allow for the variation of gravity with the latitude

$\times \left(1 + \frac{x + 52251^* + 2s}{20{\cdot}89 \times 10^6}\right)$ to allow for the diminution of gravity with height.

Loomis' Table I gives the values of 60159 log H − 27541 for values of H from 11 to 31 inches. He gives no explanation of the reason for the choice of this constant 27541, and its significance is not obvious. He remarks merely that it does not change the difference of the two quantities taken from the table.

We enter Table I with the two quantities H and H', and take their difference. This gives a first approximation to x, the difference of height between the two stations.

Table II gives the values of 2·341 $(T - T')$. It should be noted that this correction is required only when mercurial barometers are used, which is seldom. Aneroid barometers are mechanically compensated for their temperature, and no correction is then required on this account.

The resulting difference of height is then multiplied by the factor

$$\left(1 + \frac{t + t' - 64}{900}\right),$$

no table being provided for this process. The result is a close approximation to the final result.

Table III gives the value of $x \times 0{\cdot}00265 \cos 2\lambda$, by a table of double entry with arguments x and λ.

The correction for the variation of gravity with the height is split into two parts.

Table IV gives the correction equivalent to multiplication by the factor $1 + \frac{x + 52251}{20{\cdot}89 \times 10^6}$: the part involving the difference of heights. This is a table of single entry with argument x.

* $52251 = 60159 \times 2\mu$.

Table V gives the correction equivalent to the remainder of the factor $\frac{2s}{a}$. But since s is not necessarily known, though to the approximation required it may be deduced from the barometer reading at the lower station, the correction is arranged in a table of double entry with arguments H and x. It may be noted that this table is unnecessarily extended. If x, the difference of heights, is 25,000 feet, the barometer at the lower station can hardly be so low as 16 inches, corresponding to an elevation of about 16,000 feet.

The above is the form of the tables as given by Loomis, and reproduced without comment in English books up to the present day. But it should be remarked that they are slightly erroneous in Table IV. The factor

$$1+\frac{x+52251}{20{\cdot}89\times10^6}$$

is required in the reduction of observations made with mercurial barometers. But the aneroid is equivalent to a spring balance, which in itself is independent of variations in the intensity of gravity. A reference to the theory shows that when aneroid barometers are used, the term 52251 should be omitted; and this is the greater part of the correction tabulated in Table IV. We shall avoid this error if we omit altogether the correction from Table IV, and enter Table V with the mean of the barometer heights at the two stations, instead of with the barometer height at the lower station.

We may note also that as Loomis' tables are commonly printed, the argument at the side of Tables IV and V in the column headed "height" is misleading. The column should be headed "difference of height of the two stations".

The tables in the form given by Baily are a modification of the above. Baily takes as an average case that s, the height of the lower station, is 4000 feet, and that x, the difference in height of the two stations, is about 3000 feet (though he does not say so in the latter case). With these assumptions, the tables are shortened; but they are arranged so that the computation must be done by logarithms, which is less convenient for the traveller. It does not appear, then, that the form given by Baily has any advantage over the form given by Loomis; and the results are not quite so accurate in theory, because of the assumptions indicated above.

Both Loomis and Baily neglect the variation in the amount of moisture in the air, and are content to arrange their constants to correspond to an average amount of moisture. The recent tables by M. Angot, as now published in the *Annuaire du Bureau des Longitudes*, give means of allowing directly for the moisture of the air, and are in this respect a great improvement on the older tables.

But it may very well be doubted if it is of much avail to take account of refinements such as the variations of gravity with height and latitude, and variations of the aqueous vapour present in the air, while the crude, though inevitable, assumption is made that the average temperature of the column of air between the two stations is the mean of the temperatures at those stations. This cannot be exact; and it will be seen that the error introduced by this assumption may very well be much greater than the small corrections due to the other causes.

The tables for the calculation of barometer heights.

The following brief table is not sufficiently extended to be convenient in the actual calculation of observations. It is given here only that we may have an example of the style of the principal table, and of the magnitude of the quantities involved. For the reasons given above, we have not thought it necessary to give any table of the small corrections whose effect is trifling compared with the uncertainty of the principal temperature correction.

Barometer in inches	Feet	Difference for 0·1 inch	Barometer in inches	Feet	Difference for 0·1 inch
12·0	3670	217	22·0	19506	119
13·0	5761	201	23·0	20668	114
14·0	7698	187	24·0	21780	109
15·0	9500	174	25·0	22846	104
16·0	11186	164	26·0	23871	100
17·0	12770	154	27·0	24857	97
18·0	14264	145	28·0	25807	93
19·0	15676	137	29·0	26724	90
20·0	17016	130	30·0	27610	87
21·0	18291	124	31·0	28466	84

Take the difference of the heights corresponding in the above table to the barometer at the upper and lower stations.

To correct for temperature, multiply this difference by

$\frac{1}{900}$ (sum of air temperatures at the two stations $-64°$).

The boiling-point thermometer.

The pressure of the air affects the temperature at which water boils, and a determination of the boiling point of water thus affords an independent determination of the pressure of the atmosphere, and gives the same kind of limited information on the height above sea as is given by the aneroid barometer. Since it is not possible in the field to carry thermometers which can be read to less than one-tenth of a degree Fahrenheit, and one-tenth of a degree in the boiling point corresponds to a difference of pressure which is in its turn equivalent to a difference in height of about fifty feet, it follows that the boiling-point thermometer is less sensitive than the aneroid for determining differences of height. But on the other hand, it is less likely to become deranged, and it is therefore well to carry both instruments on a journey, and to use the boiling-point apparatus to control the general accuracy of the barometer.

The following is an abbreviation of the usual table for the relation between the boiling point, the barometer, and the difference of height between stations.

Boiling-point	Equivalent barometer	Height in feet above the point at which water boils at 212°
212° Fahr.	29·921	0
210	28·746	1046
208	27·613	2097
206	26·521	3151
204	25·466	4210
202	24·447	5278
200	23·461	6354
198	22·507	7439
196	21·584	8533
194	20·690	9638
192	19·828	10750
190	18·998	11867
188	18·199	12988
186	17·426	14124
184	16·681	15266
182	15·964	16412
180	15·275	17567
178	14·611	18728
176	13·970	19897

A second table is usually added, giving the factor by which the above differences should be multiplied to allow for the mean temperature of the intervening air. This is based on the assumption that the temperature at the upper station only is observed, and that the mean temperature may be derived from the approximate law that it decreases 1° F. for every 330 feet of ascent.

It appears that when the temperature at the lower station is observed, or can be estimated approximately, it is better to correct for temperature as in the calculation of barometer heights.

In a collection of survey tables it is usual to find the tables for the reduction of boiling-point observations given in the above form; and no allowance is made for the small corrections due to change of gravity in different latitudes, or for different altitudes of the lower station. These are small compared with the uncertainties

Plate IX

1. *Aneroid Barometer.*

2. *Boiling Point Apparatus or Hypsometer.*

Plate X

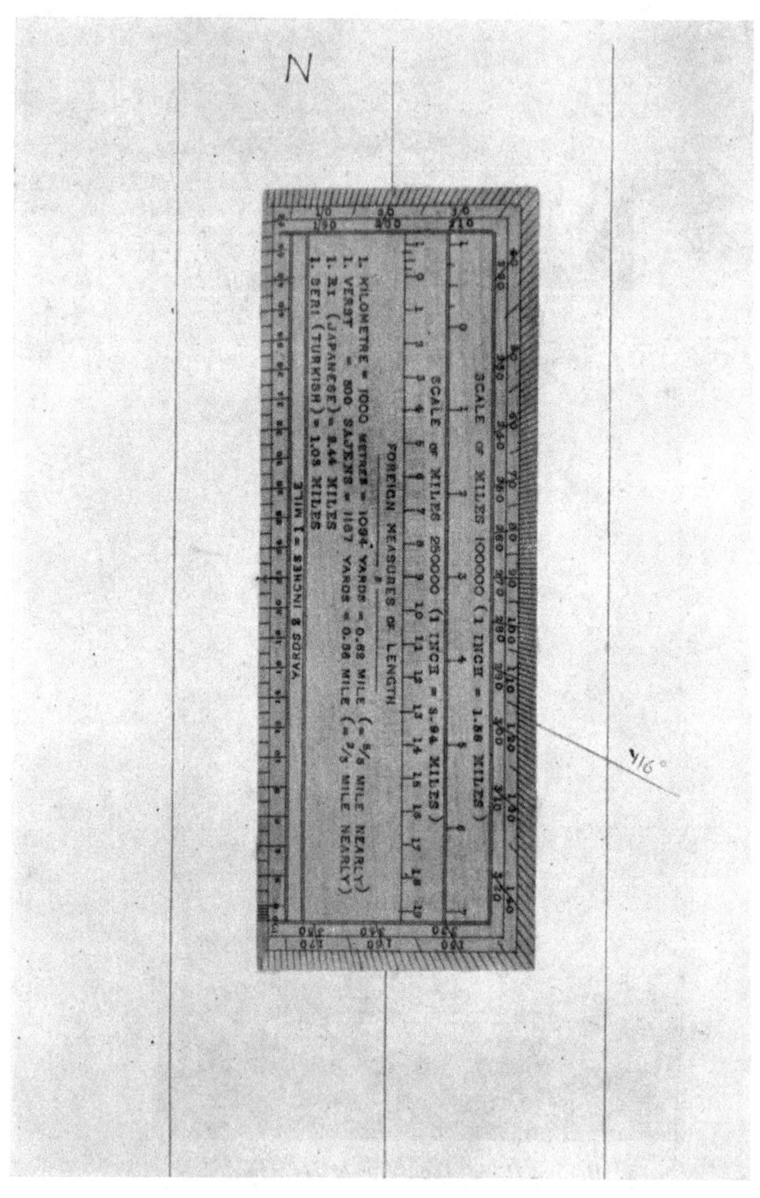

The Service Protractor set on sketching board.

of the thermometer reading. But it is well to note that the most correct way of reducing the observations is to translate the boiling points to the corresponding barometer heights, by tables such as that given above, only more extended, and to complete the calculation as for aneroid readings.

Instrumental precautions.

The principal point to attend to is that the bulb of the thermometer must not be allowed to dip into the water which is being boiled, because any impurity present in the water alters the temperature at which it boils. The bulb should be suspended so that it is fully exposed to the steam which is coming off, but is just clear of the liquid itself.

It is of course essential that the errors of the thermometers employed should be verified at the National Physical Laboratory before the start of the expedition, and again on its return. The thermometer tends to lower its zero point for a long time after manufacture, at a gradually decreasing rate. Old thermometers, well standardised, are therefore more trustworthy than new ones.

The sketch-map.

So far we have supposed the explorer on the march daily, and have considered what he can do to make a record of the route without seriously interrupting the march of the caravan. If he can halt and spend some days in one place he has the opportunity of making a more complete reconnaissance of the neighbourhood, with a sketch-map by prismatic compass or plane table. It is a map, in the sense that it covers all the area with a uniform degree of thoroughness, and does not merely draw a line across the country; and it is a map also, in the sense that it aims at representing the whole topography of the ground, the relief as well as the plan. It is a sketch, in the sense that it is made with simple and portable instruments, rapidly, and with strictly limited means; and it has no claim to great precision.

Exploratory mapping of this kind has many uses. In warfare it serves to produce temporary maps which shall fulfil all the purposes of a regular topographical map far in advance of the possibilities of a regular survey: such, for example, was the map of Burma made in 1885 by two officers of the Royal Engineers at the time of the expedition to Mandalay. In peace it may serve as a necessary preliminary to more detailed operations of survey, or to illustrate the work of a scientific expedition making a thorough study of an unmapped region, in geology, archaeology, or what not. And finally, in the teaching of geography it is admirably adapted to serve as a means of instruction in mapping which is within the powers of students, and can be executed in a limited time, with relatively simple instruments.

Such a topographical sketch may be made with the prismatic compass and a simple instrument for measuring slopes, such as the Watkin clinometer. Or it can be made with much greater accuracy, but at the expense of heavier and more costly tools, with the plane table and a clinometer of the pattern used on the Survey of India, and known familiarly as the Indian clino. In all the operations of survey the accuracy obtainable is limited by the time which can be given to the work, and the amount of baggage which it is possible to carry with the party. It is of great importance to have a clear appreciation of this principle. Much time may be saved by careful consideration before starting, of the degree of accuracy aimed at, and of the precise equipment necessary to obtain this in the most economical way. And much time is wasted by aiming at minute accuracy with rough and rapid methods, on the one hand; or by using for what should be rapid work an instrumental equipment more suited to a higher degree of deliberation and precision.

Whatever the instrumental outfit adopted, the same principle runs through all sketching, that a framework must be constructed by triangulation from a measured base, and that all detail must be hung upon this framework. If the position of one point of the triangulation can be determined in latitude and longitude by astronomical methods, such as we have already sketched in this chapter, so much the better; and again, if the sketch can be orientated correctly by an azimuth determined astronomically, again so much the better. But such determinations should be regarded as outside the limits of the methods of field sketching themselves. They add an excellent finish to the result, but they are not an essential part of the process.

Triangulation.

The framework of the sketch is built of triangles so that it shall be rigid. In order that it shall be of predetermined scale the framework of triangles is built upon a measured side. An error in the length of the base affects the scale but not the shape, and the former is more easily rectified than the latter when the time comes to fit the work to that of others. The strongest triangles are equilateral, and in laying out the framework the stations are selected to approach this shape: or at least to avoid very acute angles. The points to be occupied, called the "ruling points" in sketching, and "trig. points" in trigonometrical survey, must be in commanding positions and intervisible; they must be marked by easily recognisable objects such as church towers, stacks, single trees of peculiar shape, etc., in sketching, and suitable for the erection of beacons in more deliberate survey. The principles are much the same in all grades of triangulation. And the more time one can spend in refinement the better is the result. It may however be waste of time to aim at much accuracy, and the best surveyor is the man who adapts his methods most economically to the desired end. There is an old story of an officer who was ordered to make a sketch of the operations of the siege of Delhi, and when asked for his results after the capture of the city replied that there were some preliminary questions connected with the measurement of the base which had not yet been solved.

Compass sketching.

The instruments are

(1) The prismatic compass, of which the military pattern is by far the best, and also the most expensive, because it is fitted with arrangements for marching by night, which are not essential in work by day.

(2) The protractor, a graduated scale by means of which the observed angles are laid down on the sketching board, and distances are scaled off. Again, the military pattern, as described in the *Manual of Map Reading and Field Sketching*, is the best, and more expensive than simpler patterns, which can be used, though not so conveniently.

(3) A board on which drawing paper can be mounted, and which should have some kind of waterproof cover to protect the work from damage by rain.

A piece of three-ply wood about thirteen inches by ten, with a piece of waterproof sheeting fixed to the back, that can be drawn over the front and fastened, makes a sketching board of convenient size that can be held against the body and manipulated without need of any kind of support.

(4) A good pencil, of the degree of hardness 2H or 3H, which will take and preserve a fine point like a needle, and which should be protected in the pocket by a point protector.

It may not be superfluous to observe, for the benefit of the beginner, that good quality in paper and pencil is essential to success. The paper has to stand a great deal of wear and much rubbing out; and it will often get damp. At the end of work in the field it must be in condition to be cleaned up, inked in, and perhaps coloured. Only good drawing paper will be in this good condition when the sketch is finished; inferior paper will have gone to pieces. Nor can any good work be done unless the pencil will take and keep a fine point; and inferior pencils are the cause of endless waste of time and inaccuracy in work. Therefore do not grudge the few pence that will buy the best paper and pencils instead of the very inferior stuff in ordinary use.

The compass.

It would be tedious to describe in detail the features of an instrument which cannot be understood until the instrument itself is taken in the hand, but which then become almost self-obvious. We will confine ourselves to some general remarks on compasses and their use.

The compass card must be graduated right round from 0° to 360°. Any other method of division is almost useless. The readings increase as one turns from magnetic north, reading zero, through east to south, and round again by west to north.

It is essential that the compass, when not in use, should have the card raised off the point by the catch fitted for this purpose, or the card will rattle about, and the point of the needle on which it rests will be blunted.

To try if the point is in good condition, open the cover out flat and lay the compass on a level and smooth table. When the card has come to rest, turn the compass steadily in a horizontal plane, and see that the card remains at rest though the case is turned. If the point is good the card will not drag after the case but will remain almost unmoved.

The prism, with one face ground into a lens, allows the eye to see at the same time the graduated compass card, and the distant object. To provide for differences in the focal length of different eyes, the prism is mounted on a slide, so that the card can be brought into clear focus for any eye. Place the compass at the edge of a level table, and draw out the prism to the end of its slide. Look in at the compass card, and gradually depress the prism until the divisions of the card are seen perfectly distinctly. Note the position of the prism in its slide, and if the compass is your own property make a mark to show where the prism should be placed without having to redetermine it on another occasion.

The modern prismatic compasses filled with liquid, to damp the vibration of the card, and fitted with radium illuminated pointers and illuminated scales, for night work, are greatly improved instruments.

In the use of the compass it is necessary to acquire the trick of seeing comfortably at the same time the distant object, the line in the cover of the compass superposed on it, and the compass card mingling with it, so that one may read the degrees of the card to which the line points.

It is possible to see at the same time the distant object and the line close by, because they are viewed through the slit above the prism; this slit restricts the pencils of light entering the eye to pencils of narrow angle, so that the eye can focus on them readily enough, though they come from objects at such different distances.

It is possible to see the wire coming down over the compass card, because the view slit crosses the pupil of the eye from top to bottom. The light from the distant object enters the upper part of the pupil, and forms an image on the retina. The light from the compass card, reflected by the prism, enters the lower part of the pupil, and also makes an image on the retina. Across the centre of the field of view, where the illumination from the two sources is fairly equal, the two images are visibly superposed, and one may see the wire as if it were actually coming down and cutting the compass card. To get the right effect, it is necessary that the two images should be nearly of the same brightness. Their relative brightness can be altered by moving the eye up and down along the slit, so that more or less of the pupil is exposed to the light from the object, and less or more to that from the card. The compass card, or dial, is engraved on mother-of-pearl in the best compasses, because that reflects a great deal of light, and gives an image of the card as bright as the image of the distant landscape, which is often very bright indeed.

It is very common to see a beginner with the compass tilting it forward and downward, in the effort to see the two images plainly at the same time. This has the effect of dropping the card away from the prism, and putting the scale divisions out of focus. When one has determined the proper position for the prism, as explained above, the loss of focus is a plain indication that the compass is not being held level, which is dangerous, because the card may foul and give a quite wrong reading. Therefore it is well to be very sure about the proper position of the prism, and use this control over the tilt of the compass.

Avoidance of local deflections.

It is a commonplace that the compass becomes deranged and gives false readings if there is a mass of iron anywhere in the immediate neighbourhood. Yet it is very common to see beginners with the compass leaning against a bicycle or an iron gate, or taking bearings from the point of vantage that a bridge over a railway gives. Such mistakes are, of course, soon realised and avoided; but it is not so easy to guard against less obvious sources of error, such as a water main under the road.

Should there be any suspicion that abnormal attraction exists at the spot from which bearings are taken, it is easy to walk directly towards the objective, and repeat the bearing observation twenty or thirty yards further on. If there was local attraction at the first point, it will probably be different at the second, and the error will be detected.

These remarks apply to strictly local attractions, such as are caused by fairly small masses of iron at close quarters. In some countries, such as South Africa, the whole ground is full of magnetic rocks, and the compass is practically useless.

Compass triangulation.

In compass triangulation the angles of the triangles are not determined as such, but as the differences between the observed magnetic bearings of the respective sides. Each side of a triangle is drawn as a line making a certain observed angle with the magnetic meridian, and is subject to the errors which are inseparable from these determinations of magnetic bearing. Hence we may expect errors of at least half a degree to occur with frequency. And a deviation of half a degree is equivalent to a shift of one-twentieth of an inch at a distance of six inches, which is very much larger than the uncertainty of drawing. Hence we must realise that compass sketching is far from being an exact process, and must avoid putting too much time into a method which is incapable of giving anything like accuracy. Its merit is that it is quick, and that the instruments required are easy to carry about.

Plotting the bearings.

The observed magnetic bearings are plotted with the protractor. The paper is ruled with parallel lines to serve as magnetic meridians, and one end of these lines is marked North. To protract a given bearing from any point the protractor is placed with its centre on the point, and its long side parallel to the meridians as drawn on the paper. If the angle is between 0° and 180° the ray will lie to the east, and the protractor will be placed so that it lies east of its centre. (*Note*: the centre of the protractor is the centre from which the angular divisions are struck, and is marked by a small arrow pointing to one edge.) If the angle is between 180° and 360° the protractor is put down to the west of the centre, and an inner line of figures is used. All this is complicated to describe, but it may be learned by inspection of the instrument without any difficulty. To save mistakes, one should always ask oneself, Where roughly is the ray? If this is done, the protractor cannot be placed in the wrong position.

The base.

The necessary qualifications for a base are that it should be possible to see plenty of surrounding points from each end; and that it should be practicable to measure roughly from one end to the other. Generally speaking, one has little opportunity for picking and choosing in selecting the base for a compass sketch. The sketch must be made without delay, and it is necessary to start without an elaborate search for a good base, which is always hard to find. The length of the base may be determined by pacing, or by measurement with a calibrated bicycle wheel. And it should be remembered that even if the base is ten per cent. wrong in length, the effect on the sketch is only that it is increased or diminished in scale; it is not distorted, and if an opportunity occurs it is easy to make a new determination of some length and draw a new scale for it.

In pacing a base, it is often necessary to make some allowance for curvature of the way which the observer is compelled to go from one end to the other. Care should be taken that the correction is not over-estimated. On looking along a road, the divergences to right or left look very much more important than they really are. Divergences of fifty yards will make a road look very crooked, but they have very little effect upon the length of a mile of road between two points. Two or three per cent. reduction will be enough to allow for apparently quite considerable deviations.

The measured length of the base is laid off by means of the scale on the protractor, giving hundreds of yards on the scale of two inches to the mile.

The process of sketching.

Select the base. Occupy one end of it. Take the bearing of the other end, and plot it. Then take and plot the bearings of any points round about that seem to be suitable for ruling points. Be sure that plenty of these points are taken at the start, for as the work proceeds one will find that some of the chosen points prove to be unsuitable, and must drop out of use. Unless plenty have been taken to start with, there is danger that too few may be left.

Pace from one end of the base to the other, and scale off the distance from the protractor, making if necessary a suitable deduction for crooked pacing and slope. The further end of the base is now fixed. Take bearings to as many of the ruling points as possible. At least two points should now be determined, one on each side of the base. Proceed to these, and continue the process of observing and plotting the rays to the ruling points.

When the framework is thus built up, the second stage of the process begins. There will be important points, say at cross roads, which have not been fixed by intersections; they will now be fixed by resection, that is to say, by taking bearings from each to two ruling points already fixed, and drawing rays back on those bearings until they intersect. The point of intersection will evidently fix the point under occupation.

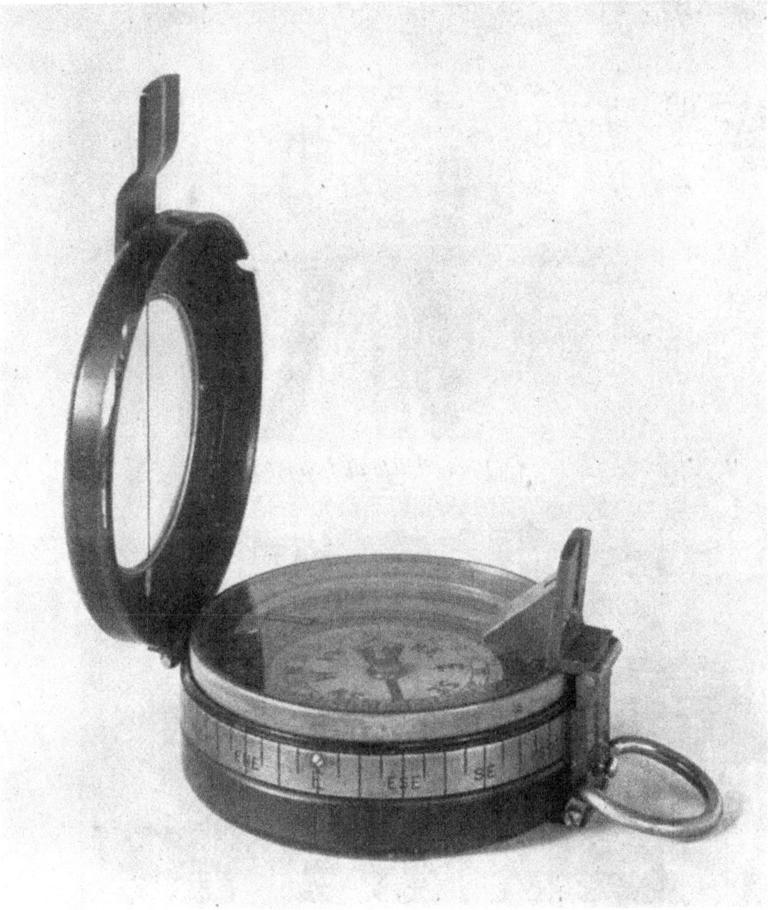

1. *Prismatic Compass, Mark VIII.*

2. *Abney Level.*

Plate XII

1. *Plane Table and Sight Rule.*

2. *Indian Clinometer on Plane Table.*

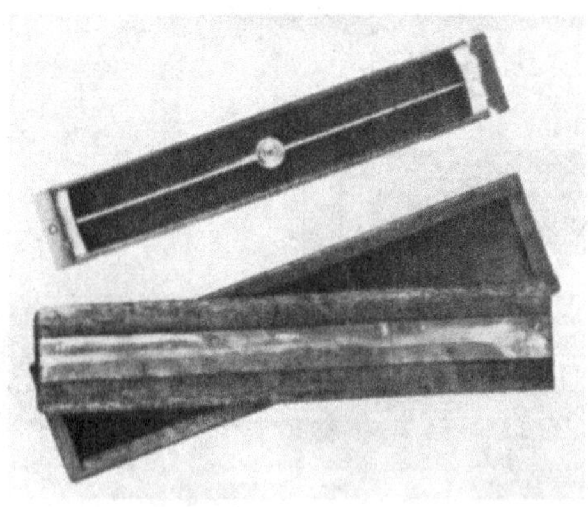

3. *Trough Compass and Case.*

Round about the intersected and resected points, the detail is sketched in by eye estimation. This will not be very accurate, but it will be rapid, which is the most important consideration. And error cannot accumulate to a serious extent, because the whole is controlled by the triangulation. The accomplished sketcher learns to economise in the use of instruments, and after the triangulation is done, he relies principally upon estimation for his intermediate detail. Also, he does not wait till the triangulation is done to begin the detail, but puts in as much as possible around each point that he occupies.

Facility in sketching cannot be taught except in the field, and success depends upon cultivating an eye for country, so that the greatest possible economy of means may be practised. But it is possible to show certain ways of economising.

For example, if it is important that a given road should be fixed as accurately as possible, it is sufficient to fix every alternate angle by resection, and to put in the intermediate angles by rays drawn down the road. Or again, if a village is to be sketched, it is usually wasteful of time to work right through it from one side to the other. Instead of doing so, work round the outside of the village, fixing points on the roads that approach it, and the directions of those roads. Having laid down the roads leading to the village, it will generally be easy to sketch the whole of the necessary detail of the village itself without being obliged to make any instrumental fixings inside. Many time-saving methods such as this will be learned by experience.

The Watkin clinometer and the Abney level.

These instruments measure the angle of slope. As with the compass, it is tedious and unnecessary to describe either instrument minutely, since its use is almost self-evident when the instrument is taken in the hand. (See Plates XI, XII.) But a few principles that are not quite evident may be examined here.

In the first place, remember that an error in the zero of the clinometer is fatal to success in its use. If the compass has an error of zero, the whole sketch is slewed round by that amount, but no further harm is done. But if the clinometer has such an error, that is to say, if the clinometer reads angles of elevation too small, and angles of depression too large by the same amount, the effect on the differences of height measured with the instrument is very serious. The difference of height between the top and the bottom of a hill would seem to depend upon whether the observations were made from the top or the bottom.

Hence the state of adjustment of the clinometer should be examined every day that it is used, for the process takes only a minute or two. Observe the same ray from opposite ends. Suppose that it reads Elevation 2° at one end, and Depression 3° at the other. Then evidently the clinometer reads half a degree low, and a correction to allow for this must be applied to all the readings made with it. It is better to determine this correction and apply it mentally, than to try and adjust the instrument by opening it up and turning the adjusting screw. This is a tedious process; and it is apt after a little to leave the screw loose, so that the error varies with every jar that the instrument receives.

The clinometer measures the elevation or depression of one point as seen from another, in angle. To convert this into difference of height in feet we must know the distance between the two points in yards. This is taken from the sketch by means of the scale of yards on the protractor. We then apply the rule

$$\text{Difference of height in feet} = \begin{cases} \text{horizontal distance in yards,} \\ \text{divided by twenty, and} \\ \text{multiplied by the slope} \\ \text{in degrees.} \end{cases}$$

The proof of this rule is very simple. A slope of one degree is equivalent to a rise of one foot vertically in a distance of 57·3 feet horizontally. Since 57·3 is an awkward number to deal with mentally, we substitute 60, which is near enough to correspond to the accuracy of the whole process. The 60 becomes 20 because the vertical intervals are measured in feet, while the horizontal distances are measured in yards, by old-standing tradition. The rule now becomes evident.

It is easy to see that the process is the more accurate the shorter the distances involved. Suppose the slope is one and a half degrees, and the distance 2460 yards. The vertical interval is 2460 divided by 20, and multiplied by 1·5, or 185 feet. But suppose that the slope had been one and three-quarter degrees; the vertical interval would then have been 215 feet. A quarter of a degree, which is less than the instrument can give with certainty, makes a difference of 30 feet in the result, and this difference is evidently proportional to the horizontal distance between the two points. At 1000 yards a quarter of a degree makes a difference of twelve feet, so that it is useless to expect to obtain heights with the clinometer which are correct within a few feet. In gently undulating country, where the distance from point to point will tend to be large, while the uncertainty in the measure of the slope is an important part of the whole quantity, the results given by the clinometer are very likely to be inconsistent and confusing.

The Abney level is more accurate but more difficult to handle: it requires some kind of support to the hand, such as a staff.

Assumed datum height.

In making a compass sketch it is not commonly the case that the height of any point above sea level is known. It is then necessary to make an intelligent assumption of some height to start with. This is called the assumed datum, or given height; and all others are reckoned from it as zero. All figures of heights entered on the sketch should be referred to this datum; they should never be entered as differences from some other point, but always as heights above sea level or datum, with the assumed datum.

Sometimes it is convenient to determine the height of the top of a tower which may be visible though the base is not. Such heights should be entered in a list on the edge of the sheet; no figure, except the height above datum of the ground itself, should ever be written alongside the object on the sheet.

Contouring with the clinometer.

The most ready method of showing the relief of the ground is by sketching the contours or form lines: the latter being the rougher and more sketchy attempts at the former. Since time is always of the greatest importance in this class of work, we must be careful that we limit the use of the instrument to the smallest possible number of observations, and that these observations are so disposed that they produce the greatest effect.

The first step is to obtain the heights above sea, or above datum, of the ruling points; this gives a framework of spot heights upon which to construct the contours. We have already seen how the height of one point is found when the height of some other visible point is known.

The second part of the process is to find how the contours lie round about one of these spot heights. Suppose that the ground slopes away uniformly in a given direction, at a slope of two degrees. What is the interval in yards on this slope between contours having a given vertical interval? The question is easily solved by inverting the relation given above, and writing it

Horizontal interval in yards = {vertical interval in feet, multiplied by twenty and divided by the number of degrees in the slope.

Thus if the vertical interval adopted for the contours is 25 feet, on a slope of two degrees the contours are spaced at a distance of 250 yards apart.

Suppose then that the spot height is 284 feet. Down the slope the first contour is that for 275 feet. Here the vertical interval from the spot height to the first contour is 9 feet, and its horizontal distance is therefore nine twenty-fifths of 250 yards, or 90 yards. Take the protractor and scale off 90 yards along the line which marks the direction in which the slope has been measured. This brings us down to the 275 contour; thence scaling off successive distances of 250 yards we obtain points on the 250, 225, 200, and succeeding contours, so long as the slope remains uniform.

Repeating this process in a different direction we obtain other points on the same contours; and these are eventually joined up by sketching.

It is clear that this process may be very wasteful of time unless care is taken that the lines of points so determined are dominant in the construction of the contours. It is almost impossible to lay down in a book the principles of economy in contouring which can be learned only by practice in the field. But one general rule is clearly useful. Run these lines of points, or contour ranges, as we may call them, along the ridges and the valley bottoms. With a range along each crest and one up the floor of the valley it is possible to sketch the whole of the valley contours without the possibility of going far wrong.

Beginners are apt to find this process of contouring with a clinometer somewhat confusing, and are to be seen working out the necessary small calculations on paper. This should never be required. The student should train himself to do all the calculation mentally, always remembering that the process is at best only a rough one, and that quantities of a foot or two have no real significance. Otherwise it would not be legitimate to use the convenient whole number 20 in place of the more accurate 19·1.

The older pattern Service protractor has scales which show the horizontal distances between contours for each degree of slope on two different scales. These are not of much use, since it is difficult to interpolate for the fractions of degrees. It is far better to accustom oneself from the start to work out mentally the distances in yards and to take these off the scales of yards found on the protractor.

Height above ground of the observer's eye.

In strictness it should be necessary to take account of the fact that the observations are made from a point about five feet above ground. On short rays one may take sufficient account of this by observing to some object such as a bush or the top of the hedge, which is judged to be of about the same height as the observer. On long rays the effect of neglecting this precaution becomes inappreciable.

Contouring by aneroid.

It is clear that the aneroid barometer can be used to determine heights, especially relative heights, and so contribute to the sketch contours. In open undulating country it cannot compete in speed with the clinometer; but it is often useful in confined situations, such as wooded gorges, from which no ruling points can be seen.

So much for compass sketching, which is sometimes useful in actual exploration, and is invaluable to the student as an introduction to the principles of triangulation and contouring. But a compass sketch is essentially inaccurate, and whenever possible the plane table should be used instead.

The plane table.

The plane table is unique among survey instruments in that it enables the surveyor to draw a map without measuring any angles or doing any numerical work, except in the contouring. The plane table is, in fact, a drawing instrument, by means of which the map is drawn in the field without the intervention of any angle-measuring instruments. It is exceedingly well adapted for rapid survey, and is very much used for making maps in a hurry, as may be necessary in military operations in an unmapped country, or during any rapid exploration.

The instrument consists of a drawing board covered smoothly with drawing paper, and mounted on a light but rigid tripod, upon which the board can turn smoothly, and can be clamped in any position desired. The accessories are:

(1) A sight rule, preferably of boxwood, having folding sights

which can be turned up at each end. One sight has a narrow vertical slit in it; the other consists of a vertical wire stretched across an open frame.

(2) A trough compass, which is a long compass needle mounted in a narrow box, with a short scale at each end. When the needle is pointing to the centre of the scale at each end it is parallel to the sides of the box.

(3) A hard, well-pointed pencil of good quality.

The sight rule should be graduated along one edge in inches, and along the other with a scale of yards corresponding to the scale on which it is proposed to work: very frequently the scale of two inches to one mile.

This is the description of the simple plane-table equipment which is generally to be preferred in exploratory work. The instrument may be complicated by fitting a telescope to the sight rule; by making the sight rule as a parallel ruler; by fitting levelling screws to the tripod; and in many other ways. It is generally unsound to try to convert the plane table into a poor kind of theodolite, but see later p. 199.

The plane table has two principal uses, which must be distinguished one from the other. It can be used to produce a complete map, entirely by its own resources, without the use of any other instrument than the accessories which normally accompany it; the triangulation can be made with it, and the whole detail filled in with it. Or it can be used to fill in the detail after the triangulation has been made with the theodolite. The first is the use to which it is put in rapid and exploratory mapping, with which we are now concerned.

Graphical plane tabling.

By graphical we mean that the plane table is to be used as a drawing instrument, to make the complete map without other instrumental aid. Both triangulation and detail are to be done with it.

The principles of the triangulation are naturally the same as those which we have already considered in compass sketching, and it will not be necessary to repeat that the process consists of measuring a base, and building up on it a framework of well proportioned triangles.

Measurement of the base.

The base must be chosen in as open and level a piece of country as can be found, and its ends must be marked in some conspicuous way, so that they are visible one from the other, and from the surrounding points which will be occupied in succession to extend the base and start the triangulation. Plane tabling is a much more accurate process than compass sketching, and a correspondingly greater degree of care is required in choosing the ground for the base.

Let us take two solitary trees or posts on a straight open road, being careful to see that they can be identified with certainty from the surrounding country, and not confused with neighbouring trees or posts. We shall get the base as long as possible: perhaps from half a mile to a mile long.

We have now to measure the base, and the more accurately the better. If chain or steel tape is available, it should be used. Failing these, a good deal can be done with a carefully calibrated bicycle wheel; and failing any kind of measuring instrument the base must be paced. The error of the result may in this case be three or four per cent. But this affects only the scale of the map, not the relative configuration of points marked upon it. And for many purposes the exact scale of the map is not very important, so long as the topography is correct.

Plotting the base on the plane-table sheet.

The base is measured in yards, and the length corresponding, on the scale of the intended map, is taken from the scale of yards on the edge of the sight rule. Consider the position of the base in the area which is to be mapped, whether it is to the centre or to the side; draw a fine line with the sight rule in the convenient position, and lay off on it the length representing the base measure, making fine perforations in the paper with the sharp hard point of the pencil, so that other rays may be passed accurately through these points when required. It is well to draw continuations of the line on which the base is to be laid off, at each end of the sight rule, so that the rule may be laid down accurately along its original direction in a subsequent step of the process; usually the base line is so short that the rule cannot be laid down on it again with any great precision.

Call the ends of the base A and B.

Set up the plane table at A, as close as possible to the mark. In plane tabling on the scale of two inches to one mile, one-hundredth of an inch represents about nine yards on the ground, so that divergences of a yard or two from the exact position of the station are hardly visible upon the map, being within the limits of error in drawing. Lay the sight rule along the line of the base on the table, and turn the table till the line of sight falls on B; then clamp the table. Beginners find some difficulty in sighting along the rule, and are seen looking sideways in very awkward positions. It may be helpful to say that the correct position for plane tabling is as nearly as possible the correct position for wicket-keeping in cricket.

Beginning the triangulation.

The last process has set the plane table. That is to say, the base line drawn on the table is parallel to the base line on the ground. And the point A on the table is over the corresponding point on the ground. We can therefore draw the ray from A to any other point by placing the sight rule so that its edge passes through A on the table and is directed to the point C on the ground. A convenient way of manipulating the sight rule is to stand the pencil (hexagonal) with one corner on A, holding it with the left hand, and with the right hand swing the sight rule about this corner as a pivot until it is sighted on the point C. Then carefully rule a line passing through A towards C. It is essential that the pencil have a fine needle-like point, and that it be held carefully to make a line really parallel to the edge of the rule. It is not necessary to draw the whole line AC. Make an estimate of the distance of C; suppose it is one mile and a half. This will be three inches on the paper if we are working at two inches to one mile. Draw then a piece of the line only, about three inches from A. If the piece is not long enough it may be lengthened later. But it should be remembered that all the rays drawn must be rubbed out in the end, so that it is economy to draw as little of the ray as will serve its purpose.

In the same way draw rays to other points $D, E, F, \ldots$ which may be useful as ruling points. We leave A with the knowledge that all the points $C, D, E, F, \ldots$ which are to be mapped lie somewhere on the lines $AC, AD, \ldots$ which have been drawn; their precise positions will be determined by other rays drawn from $B, \ldots$ intersecting these first lines.

Now move the table to B, and set it by laying the sight rule along the line BA and turning the table until the rule is set on A. The table is now set; that is to say, it has been moved from A to B parallel to itself, so that any line already drawn on the table is parallel to its position on the ground. Having set the table we draw rays BC, BD,... to the points already observed from A, or to as many of them as are visible from B; and thus C, D,... are fixed.

Two points, one on each side of the base, must be fixed by the intersections of two rays only, from A and B; and it is important to see that these intersections are as nearly at right angles as possible. A good intersection is a blunt intersection. When the rays cut one another acutely a small error in either will make a good deal of difference in the place of the intersection.

From C and D we proceed to fix other points; and we must not accept any point as well fixed unless it depends on three rays which intersect in a point without visible deviation. Working on in this way we build up a framework of triangles upon the base in the same way as we did in compass sketching. But the accuracy of the result is considerably greater.

Choice of stations.

Only experience in the field can teach how to select the stations of a triangulation with advantage; and the selection will naturally be governed by the nature of the country. The stations must fulfil many conditions:

(1) they must be in commanding positions, so that stations all round are visible from them;

(2) they must be well-marked natural or artificial objects, such as a sharp summit of a hill, or the trunk of an isolated tree, or the top of a tower. The top of a tree surrounded by low growth, and the point of a steeple, are not suitable objects to select for the continuation of the triangulation, since they cannot be occupied in their turn;

(3) they must make well-conditioned triangles, with no very acute angles.

It must be possible to occupy the stations: but it is by no means necessary that the plane table should be centred accurately on the station mark. On the scale of two inches to the mile a hundredth of an inch is about nine yards, so that the table may be several yards off the centre without perceptible error resulting. If one is working on so large a scale that errors of centering or questions of the dimensions of the instrument in relation to the ground arise, then the plane table is being used unsuitably. And for this reason its use cannot be demonstrated properly in a class-room.

It will often happen that without a preliminary reconnaissance and occupation of the proposed stations, the surveyor will be deceived. A point ahead may appear to be an admirable station, but when it is occupied it may be found that further progress is obstructed by natural or artificial obstacles which could not be seen from the last station. It is therefore advisable to draw rays to more points than are really required for the triangulation. This will allow for dropping out stations which prove in the end to be unsuitable.

It is one of the great advantages of plane tabling that a great number of rays may be drawn with very little trouble, so that superfluous rays need cause no regret, and the survey can go ahead without preliminary reconnaissance. In a theodolite triangulation the case is very different. Here a reconnaissance is absolutely necessary, for selecting and beaconing the stations, and by far the best kind of reconnaissance is a plane-table triangulation.

The use of the triangulation.

When the triangulation is complete our plane-table sheet is covered with a series of rays intersecting in points, which are conspicuous points about the country, tops of hills, isolated trees, church towers, and so on. The triangulation is the skeleton of the map, but the visible map is not yet begun. It may be that we do not propose to go any further immediately. We have laid out the framework of triangles; from this we can select the particular scheme best fitted for observation with the theodolite, and we are then ready to beacon the country for the deliberate observation. But if time presses, and we wish to have an approximately complete map as quickly as possible, we must dispense with the precise triangulation, and proceed at once to fill in the detail of our plane-table sheet.

The map is to show the natural features of the country, hills, valleys, and rivers, and the artificial features, railways, roads, and towns. These, the most important topographical features, are not as a rule marked by upstanding conspicuous points suitable for triangulation stations, while, on the other hand, very excellent triangulation stations, such as isolated rocks or trees, are not features that will eventually appear at all upon the topographical map. Our skeleton, then, is no part of the map proper, and is destined to disappear in the end, leaving no sign of the rays which constructed it, and only the small triangular signs to mark the stations which were used.

Accuracy of the graphical triangulation.

The sight rule can be set upon an object with an accuracy of two or three minutes of arc; this is readily seen if we consider that the diameter of the sun or moon is about thirty minutes of arc, and that they are very large objects to set upon if we view them along the sight rule when they are near the horizon. It is not hard to set with an accuracy less than one-tenth of the diameter of the moon. On the other hand, the error of a compass bearing, observed without a tripod or other rest for the compass, is likely to be at least half a degree. In compass sketching each ray is plotted independently as an absolute magnetic bearing. In plane tabling the board is set up at each new triangulation point by sighting back with the sight rule on one of the other points already well fixed. The process depends only on the care with which the settings are made, and the stability of the table. Hence the rays drawn on the plane table are very much more accurate than those plotted from compass bearings. By increasing the bulk and weight of our apparatus we have much increased its accuracy.

Intersected points.

The stations of the triangulation are perhaps two or three miles apart. As each one is occupied we take advantage of its commanding position to draw rays to all the conspicuous objects round about —not with the idea of occupying them in turn as triangulation stations, but in order that they shall be well fixed by intersecting lines. In this way a great number of steeples, chimneys, well-marked trees, and such objects, will have been put in on the plane-table sheet. These are called Intersected points. In what follows they serve as auxiliary to the principal points of the triangulation; and like them, are destined in great part to disappear from the finished map, since they are often of no topographical importance.

The use of the trough compass.

At some point of the triangulation, when the table is orientated by setting on another triangulation point, the trough compass is taken from its case and laid on the table, with the end of the needle which is marked with a cut toward the north. The compass is turned until the needle points to the zero of the scale at each end. The needle now lies in the magnetic meridian, and the sides of the box, being parallel to the line of zeros of the scales, are also in this meridian. Draw pencil lines along the sides of the box, and mark the north end. (See Plate XII.)

Now we are able to set up the plane table in approximate orientation at any point whatever of the ground. Set up the table; take out the trough compass and place it on the compass lines drawn as above, taking care that the north end of the needle lies to the end marked North. Then turn the table until the needle points to zero. The table is now orientated as accurately as the compass permits, that is to say, within about a quarter of a degree, unless there are magnetic disturbances about.

The purpose of this orientation by compass is to facilitate the process of resection which follows.

Filling in the detail.

Up to the present we have been fixing well-marked points by drawing intersecting rays from other points previously determined. When a sufficiency of these have been fixed, we begin the new operation of putting in the important topographical detail, the roads, fords, railway bridges, and so on, which are inconspicuous objects in general, as viewed from the ruling points, and cannot be determined by intersections. But, on the other hand, the ruling points should be quite conspicuously visible from them; and if three are visible we can (with certain limitations to be discussed at a later stage) set up the plane table and determine the place on the map, by the process known as resection. Perhaps retrosection would have been the better term, since the process consists of drawing rays backwards from the ruling points. A point determined by resection is often called a plane table fixing.

Plane-table fixing by resection.

Whenever three ruling points, suitably disposed, are visible, a plane-table fixing may be made. The conditions for suitability are

(1) that the points are not too close together, neither is one nearly opposite another;

(2) that they do not lie nearly on a circle passing through the point which is to be fixed.

The reason for these restrictions will appear in due course.

By means of the trough compass set up the table in approximate orientation. Set the sight rule so that it is directed to one of the points, while its edge passes through the place of this point on the table; and draw a ray back. Do the same for the other points. If the three ruling points were correctly fixed, and the table in correct orientation, the three rays would meet in a point, which would be the point on the table corresponding to the place where the table stands on the ground.

More often than not the three rays do not meet accurately in a point, but make a small triangle, which is called the triangle of error. This is not due primarily to any error in the positions of the ruling points, nor to inaccuracy in drawing the rays, but to error in the orientation of the table as set up by the trough compass. The compass cannot be relied on to within $\frac{1}{4}$ to $\frac{1}{2}$ a degree at the best; and there may be local attraction. Moreover, owing to the convergence of the magnetic meridians, lines drawn along the magnetic meridians at different parts of the sheet will not be accurately parallel to one another. In dealing with the triangle of error we assume that the whole of the error is due to this faulty setting up of the table.

Solution of the triangle of error.

It might be supposed that when there is a triangle of error, the true place which we are trying to fix would be at the centre of gravity of the triangle. But this is entirely wrong. It is not necessarily inside the triangle at all, as will be seen from the following considerations.

Let A, B, C be our three ruling points on the map, and O the point in process of fixing. If the table is not rightly orientated it is rotated about the point where it stands on the ground, and we may represent this by turning it in our figure about the point O on the map, so that A, B, C are moved to A', B', C' on the table. Now draw the rays again. Since the real points are distant, the new rays will pass through A', B', C', and will be parallel to the old rays. They will intersect to form a triangle; but O is more likely to be outside than inside the triangle.

The student is advised to draw a few cases for himself. He will then easily understand, and be able to prove, the following rules for constructing the point *O* from the triangle of error.

Rule 1. If the table is set up within the triangle formed by the three points on the ground, the place of the table on the map will be inside the triangle of error; if not, it will be outside.

Rule 2. The point *O* will lie always to the right or always to the left of the three rays, as one looks along them to the points from which they are drawn. (It will be seen that Rule 1 is really included in Rule 2, but it is convenient in practice to state the two separately.)

Rule 3. The perpendicular distances of *O* from the three rays are proportional to the distances from the table of the corresponding points on the ground.

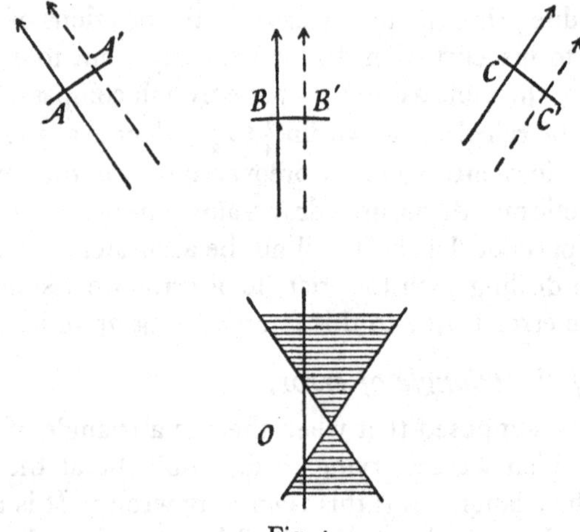

Fig. 4.

To take an example as in the figure. Let the three rays be drawn as shown, and let *O* be outside the triangle *ABC* on the ground.

By Rule 1 *O* on the map is outside the triangle of error.

By Rule 2 it cannot be in either of the sectors marked by shading.

By Rule 3, for the proportions of the perpendiculars, it cannot be in the remaining right-hand sector, but must be in the left, at the point indicated.

The method of making a plane-table fixing is complicated to explain. After a little instruction and practice in the field it is perfectly easy and rapid in execution. And it is the most ordinary operation in plane-table survey. Point by point plane-table fixings are made all along the roads, rivers, and railways; and thus the detail of the map is quickly filled in. It will be understood that only practice in the field can show how to economise effort by making the fixings at the points where they are most effective. No written explanation can give an eye for country.

Check on the solution of the triangle of error.

If the triangle of error is large, or the surveyor inexperienced, it is well to have a check on the accuracy of the solution that has been made. This is very simple. By hypothesis the triangle has arisen from bad orientation of the table. When the first solution has been made, lay the sight rule along the point which the solution gives, and one of the ruling points which has been used. Look along the sight rule, and the distant point will be off the cross wire. Unclamp the table, and turn it till the point comes on. Then clamp, and repeat the resection. This time the triangle of error should be exceedingly small, if not an exact point, and the solution should give a result very near the first solution. If it does not, something is wrong. Inspection of the solutions will probably show that an error has been made. If, however, the solutions are made according to the rules, and yet give discordant results, then it is probable that there is some error in one or more of the ruling points, and this must be put right by revisiting the triangulation points until all is verified. It is no use trying to get good results from resections until the ruling points are really laid down in their true places.

Case in which the solution fails.

The solution fails when the point which is to be fixed lies on or near the circle passing through the three chosen ruling points. If it lies exactly on the circle the solution becomes quite indeterminate. However wrong the orientation of the table may be, the three rays meet in a point, and it will appear that a perfect result has been achieved. But turn the table round into another orientation, and repeat the process. Again the three rays meet in a point, though quite a different point. This curious result depends on the well-known proposition that all angles in the same segment of a circle are equal to one another.

If the point to be fixed lies very nearly on the circle, the result is very nearly as indeterminate, and no good solution can be obtained. Great care must be taken, therefore, that the three ruling points are chosen so that there is no chance that the problem may become indeterminate in this way. If the three points in order are *A*, *B*, *C*, try to ensure that *B* is nearer than either *A* or *C*. The circle passing through them will then lie far away from the point which is occupied, and the solution will not fail.

Remarks on the part played by the compass.

It is essential that the beginner should plainly understand the part played by the compass in this operation. The compass is an uncertain instrument, and cannot be relied upon to give the orientation right within half a degree. For this reason the orientation of the plane table, when it is set up by compass for the resection, is always to be considered suspect. The size of the triangle of error is an indication of how much it is wrong; and the solution of the triangle of error eliminates completely the effect of the wrong orientation. In fact the compass is used only for convenience, in order that the triangle of error may be small. The compass is not indispensable, and if it is lost or broken the process may be worked without it. Set up the table by estimate in about the right orientation, and make a solution. The triangle of error will be very large, but it can be solved approximately. Re-orientate the table on the result of the first solution, and make a second. This will come out with a very much smaller triangle. If necessary repeat the process again, and this time the result will be satisfactory.

It is clear from this that the rôle of the compass is quite subordinate, and that any error in it will not affect the ultimate accuracy that is achieved.

Alternative method of resection with the plane table.

The following method has the advantages that it does away with the use of the compass, and that it gives (or should give) a direct solution without triangle of error to solve.

Suppose that we letter the points *A*, *B*, *C*, choosing *A* and *B* so that the station does not lie near the line joining them. The points on the ground corresponding to *A*, *B*, *C* on the table, are called A, B, C.

Lay the sight rule along *AB* (eye end at *A*) and turn the table to sight on B. Clamp.

Pivoting the sight rule on *A*, and sighting on C, draw a ray back through *A*.

Unclamp. Lay the sight rule along *BA* and turn the table to sight on A. Clamp. Pivoting on *B*, sight C and draw a ray back through *B* to intersect the ray first drawn in *C′*.

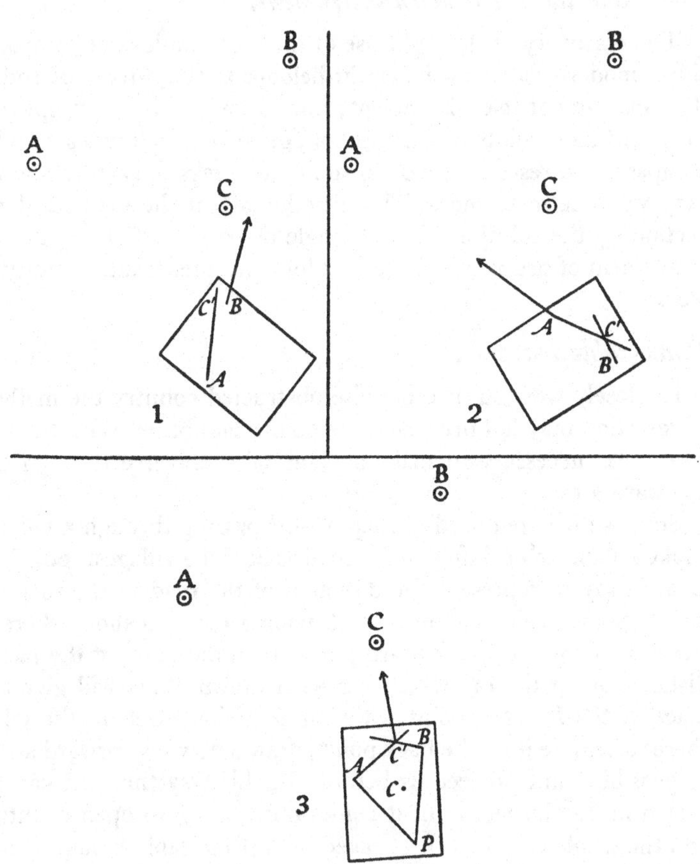

Fig. 5. *Plane tabling by direct solution.*

Unclamp. Lay the sight rule along *CC′* and turn the table to sight on C. The table is now in orientation. Pivoting in turn on *A* and *B*, sight on A and B and draw rays back to intersect in *P*, which should fall on *CC′* or this line produced. *P* is the plane-table position of the station. If *P* does not lie within half a millimetre of CC' the solution is unsatisfactory, and should be repeated with other trig. points. If the error is due to faulty map detail and not to bad drawing, make several determinations from different selections of points and take the mean.

The plane table a modern instrument.

The discovery of the right use of the plane table is of comparatively modern date, and its credit belongs to the Survey of India. The instrument itself is ancient; but so long as it was employed only for fixing points by intersection, or so long as it was set up by compass, and resections were made by two rays only, it was not a very valuable instrument. The introduction of the method of resection by the solution of the triangle of error made it at once an instrument of precision, unexcelled for convenience and rapidity of work.

Plane-table traverse.

In closely wooded or otherwise obstructed country the method of resection may fail because three points cannot be seen. In such cases it is necessary to make a plane-table traverse through the awkward area.

Suppose it is required to map a road passing through a village. Make a plane-table fixing on the road as near the village as possible. Draw a ray to represent the direction of the road to the furthest visible point. Take up the table, leaving a mark to show where it stood, and pace to the forward point; then measure off the paced distance along the ray which has been drawn. This will give the place of the forward point with fair precision. Set up the table there; orientate it on the back point; draw a new ray forward as far as possible; and proceed as before. In this way the road can be drawn in, leg by leg, until it comes out again into open country, and the whole can then be checked by a plane-table fixing. Probably the result of this will not agree with the position as carried through by the traverse, and it will be necessary to adjust the traverse to make it fit on to the plane-table fixings.

Adjustment of a traverse.

Suppose that the traverse $ABC...M$ ends at M, and that a plane-table fixing at this end makes the position M'. It is required to adjust the traverse so that it closes on *M'*. Join *MM'* and draw lines parallel to *MM'* through each angular point of the traverse. Each of these points must be moved along its corresponding parallel by a fraction of the length *MM'* proportional to the distance from *A*. This rule is of course arbitrary. It is not pretended that the result is precisely right; but some systematic method of adjustment is required, and this appears to be the best.

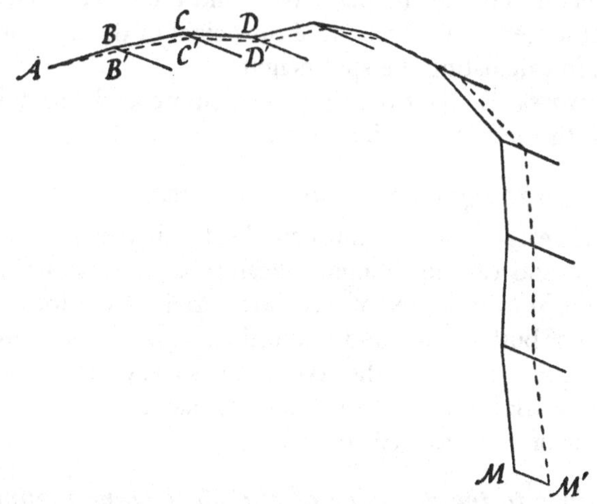

Fig. 6. *Adjustment of traverse. The full line is the original; the dotted line is the traverse adjusted to close on M'.*

In practice the rule is simplified thus: Suppose there are ten legs to the traverse, ending at *B*, *C*, *D*, ... *M*. The *B* is moved one-tenth of *MM'*, *C* is moved two-tenths, and so on, which brings *M* to *M'*; and the traverse is adjusted.

Heights and contours.

For rapid and approximate work the heights and contours can be done with the Abney level or the Watkin clinometer, precisely as in compass sketching, and it is not necessary to repeat here what we have already said on pages 147–151.

For more accurate work we may use the Indian pattern Clinometer, commonly called the Indian Clino. This consists of a brass bedplate with a bubble and levelling screw, set up upon the plane table along the ray to be observed, and levelled. Two leaves stand up at either end of the base plate. One carries a small sight hole; the other has a vertical slit, divided along one edge with a scale of degrees, and along the other with tangents of degrees, the zeros being level with the sight hole. The observer, looking through the sight hole at the distant point, reads off the tangent of the elevation or depression of the distant point. He then scales off its horizontal distance, multiplies it by the tangent, and obtains at once the vertical interval between the plane table and the distant station. He must not forget to take account of the height of the table above the ground, in calculating the spot height.

This process is about ten times as accurate as determining spot height with the Watkin clinometer.

Characteristic signs, and style of drawing.

The appearance of the finished sheet will depend very much upon skill and care in draughtsmanship, and particular attention must be paid to the conventions which are adopted for the representation of both natural and artificial features. These conventions vary to some extent with the style of the survey—for rapid reconnaissance work the style is less elaborate than in the more leisurely operations of an organised survey.

Limitations to the accuracy of graphical plane-tabling.

The method of graphical plane-tabling discards numerical measurement and computation, and relies upon pointing, generally without optical aid, and upon drawing. Every line drawn upon the table is liable to be in error by an amount on the border line of visibility; these errors accumulate until they become quite considerable, and cannot be tolerated on a map with any pretensions to accuracy.

Further, there is a difficulty to be discussed more in detail later—the question of the projection, and the effects of the Earth's curvature, which make it impossible to carry on a survey continuously through a number of sheets, and troublesome to derive latitudes and longitudes of the places mapped.

Hence plane-table triangulation has strictly limited uses. It is admirable for exploratory and reconnaissance work, but it cannot be used as the basis of a deliberate survey. This must depend upon theodolite triangulation and calculation.

But as an instrument for filling in the detail of a precise triangulation the plane table is unsurpassed, and we shall have further occasion to consider it.

Summary of exploratory methods of survey.

It will be useful to sum up briefly the methods which we have been considering, suitable for preliminary mapping, exploration, or reconnaissance.

A single-handed traveller, whose main business it is to get through the country, on exploration or on official duty, cannot make a map. But he can

(1) make a compass traverse of his route, and fill in a small amount of detail on each side;

(2) make astronomical determinations of latitude and azimuth, and with more difficulty of longitude, which fix the main points of his traverse, and serve as a general check upon it.

When he is able to halt for a few days at a time

(3) small areas of country can be sketched with the compass and clinometer;

(4) larger areas can be mapped by graphical plane-tabling.

The most successful explorer will be the man who knows precisely the advantages and possibilities of all the methods, and makes skilful use of any or all of them as circumstances may dictate.

Chapter IX

TRIGONOMETRICAL SURVEY

IN the preceding chapter we have considered the various operations of survey which may and should form part of the work of any explorer, or of any administrative officer on tour in an unmapped country. The first maps of the country will be compiled from such sketches, and will be a continual source of annoyance, because nothing will be quite right. No compilation of patchwork survey can produce a map worthy of the name, and the sooner the survey can be put upon a systematic footing, the greater will be the final saving of expense, because the less isolated work will there be to scrap. Unfortunately one argues here in a circle. A country cannot be properly administered until it is mapped, and it cannot be mapped until it has a considerable revenue to support the expense. But a new country must start its development with borrowed capital, and the truest economy is to spend some of this in survey as early as possible.

Preliminary considerations.

Before laying out the plan of our operations we must consider whether our aim is restricted to producing a map which shall have no sensible error anywhere; or whether we desire that our operations shall be of the refinement required when they are to make a contribution to geodesy, properly so called: that is to say, to the determination of the size and shape of the Earth.

For the present we will deal with the former case, and suppose that we are to undertake the topographical survey of a large isolated island, of area about 10,000 square miles. The work is to be good, but not of geodetic accuracy. The methods employed are to be as economical as is consistent with the production of a first-rate topographical map. The island is mountainous and the country fairly open, so that it does not present any difficulties of an exceptional nature. It is suitable for a good theodolite triangulation, with plane-table detail.

Simple triangulation with the theodolite.

Considered in the simplest possible way, the theodolite, as used in triangulation, consists of a graduated circle fixed in a horizontal plane; a telescope which can be rotated about a vertical axis passing through the centre of this circle; and a pointer attached to the telescope, which can be read against the circle.

We shall postpone to Chapter XIII the consideration of all the instrumental details of the theodolite, and for the moment confine our attention to the outlines of the method.

Suppose that we have three points A, B, C marked on the ground in some visible way, and that the theodolite is set up over A. It is pointed on B, and the reading of the pointer on the circle is recorded; it is then pointed on C, and the circle is read again. The difference between the two circle readings will be the angle BAC. The theodolite is then moved to B, and the angle ABC is measured; then to C, and the angle ACB is observed. These three angles should of course add up to 180°, unless the triangle is so large that the curvature of the Earth must be taken into account; and this will be a check upon the accuracy of the observations.

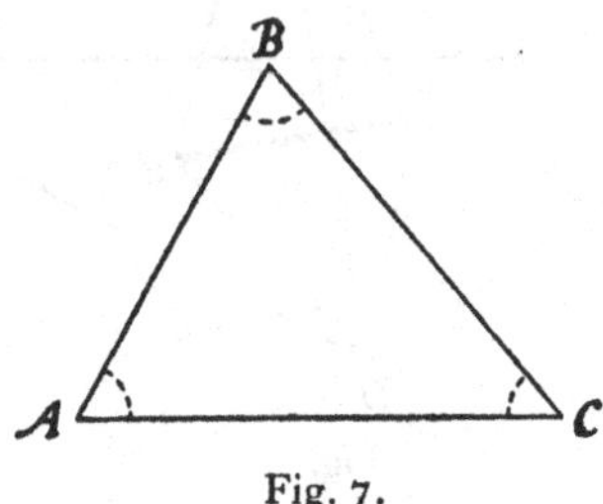

Fig. 7.

When the three angles are known the shape of the triangle is known, but not its size. To get its size we must measure the length of one of its sides; and then by a very simple piece of trigonometry we calculate the lengths of the other sides, from the known values of the angles.

In a triangulation, therefore, we must measure the length of one of the sides of one of the triangles. The lengths of all the other sides of the whole triangulation will be worked up by calculation from the observed angles and the length of the initial side, which we call the base.

Suppose, then, that we have selected six points on our ground, *A*, *B*, *C*, *D*, *E*, and *F*, disposed so as to form the triangles of our figure, of which, it will be observed, no one has a very acute angle. Each point must be visible from the other points of the triangles to which it belongs; but it is not at all necessary that all the points should be intervisible. For example, *C* must be visible from all five others; *B* from *A*, *C*, and *D*; but *E* need not be visible from *A* or *B*. These points are marked by pegs driven into the ground, and signals are erected over them.

The theodolite is set up over each peg in turn, the signal being removed if necessary, and the angles in the triangles are measured. Each triangle is checked by the necessity that the sum of its angles should differ from 180° only by the small quantities which may be allowed as errors of observation.

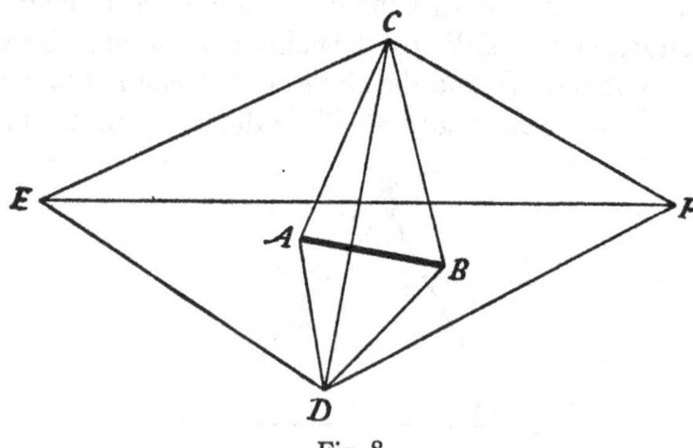

Fig. 8.

It is important to notice that, though the points of the triangulation may be at different elevations, the angles that are measured are the angles of the projection of the figure upon a level surface. For suppose that in the triangle *ABC* the point *B* is higher than *A*, and *C* higher than *B*. If we wished to measure the true angles of the triangle *ABC* we should have at each station to tilt the theodolite so that the graduated circle lay in the plane of the triangle. If, on the other hand, we keep the circle horizontal, what we actually measure is not the angle *BAC*, but the angle between the vertical planes through *AB* and *AC* respectively; which is of course equivalent to the angle at *A* in the triangle *ABC* projected on the level. But this is what we want; we have already remarked that the plan or map must represent the projection upon a level surface, not the actual configuration upon the undulating surface of the ground.

When all the angles are observed we have the shape of the figure. One side must be measured, and any one of these will do. We shall naturally select the one that lies on the most level and unobstructed ground. We suppose it is *AB* which is measured with chain or tape. Then in the triangle *ABC*, knowing all the angles and one side, we can easily calculate the lengths of *AC* and *BC*.

One of the sides of *ACD* thus becomes known, and the others follow immediately, since all the angles are known, as before.

In this way, from one measured side of one triangle only, we arrive at the lengths of all the sides of all the triangles.

The stations of the triangulation will naturally be for the most part on the summits of hills, from which an extensive view can be obtained. But it will generally not be convenient to measure from one of these stations to the next, because the intervening ground will not be open and level, and otherwise suitable for the purpose. In practice, therefore, one chooses a site for the base, and arranges a special piece of triangulation to connect it with the nearest side of the main system of triangles.

Let us look again for a moment at the reasons for making this framework of a few big triangles covering the ground which we propose to survey in detail afterwards. Suppose that our base is only a few hundred feet long, and that the country is so rough that we cannot hope to find a place where we can measure a longer length on the ground with any accuracy. It is very easy, and does not require any special precautions, to measure this base with an accuracy of one in ten thousand. If our triangles were quite exact we should arrive at the lengths of our longest sides with the same proportional accuracy. This will not be achieved in practice, but the diminution in accuracy can be controlled by the amount of care that we are willing to give to the triangulation. Suppose that it is so done that the error in the side most remote from the base is increased to an uncertainty of one in three thousand. The principal points being related to one another with this degree of accuracy it is not possible for error to accumulate over any considerable part of the plan to a greater extent than this. In cutting up the bigger triangles into smaller, either with the theodolite or with the chain, local errors may be made, but their effects are not cumulative. They cannot extend outside the particular triangle in which they are made.

Again, if only we can find points for the principal triangles which are visible from one another, the accuracy with which they are fixed is independent of the character of the intervening ground, which may be of a kind to render any sort of linear measurement impossible.

Or, to put it another way: It is very hard to measure distances accurately with chain or tape except upon open and level ground;

it is very easy, by means of the theodolite, to measure angles from one well-defined point to another. We shall expect, therefore, to do well with a method that gives us the triangles by measuring their angles, and requires that one side only shall be measured, for which a site can be chosen in the most favourable place.

With this introduction we may proceed to a more thorough consideration of the methods of trigonometrical survey, so called because we have to work by trigonometrical calculation, in place of the graphical methods heretofore employed.

The operations divide themselves into the following sections:

1. Determination of mean sea level.
2. Preliminary plane-table reconnaissance.
3. Beaconing for the triangulation.
4. Determination of a latitude, longitude, and azimuth.
5. Measurement of the bases.
6. The theodolite triangulation.
7. Determination of heights by theodolite.
8. Calculation of the triangulation and heights.
9. Transference of the triangulation points to the plane-table sheets.
10. Mapping by plane table.

Determination of mean sea level.

The zero point for heights must be the mean level of the sea on the coasts of the island. This must be determined at one or more points by the maintenance of tide gauges.

The tide gauge consists essentially of a well, connected with the open sea by a pipe of small diameter, allowing the water in the well to assume the average level of the open sea, but damping down the quick oscillations of the waves. A float in the well is connected with an indicator and scale, so that the height of the water in the well can be read at intervals; or better, it is connected with a pen drawing a trace upon a clock-driven drum, giving a continuous record which can be measured up and the results tabulated.

The selection of a site for the tide gauge is often a matter of some difficulty. We require the height of the open sea, and not the height of the water in some inlet or harbour, which is often modified by prevailing winds and currents. We require that the record shall be continuous over a long time, not interrupted by choking of the pipe with sand or weed.

The observed oscillation of the tide is the sum of a great number of oscillations of very different periods and amplitudes, the shortest periods being the halves of the solar and the lunar days, the longest the period of revolution of the Moon's nodes, nineteen years.

Further, variations of the barometer affect the height of the sea, a fall of the barometer by one inch being accompanied by a rise in the sea of about one foot.

Hence it is not possible to lay down any definite period during which tidal observations should be continued. Fifteen days' observation will cover the principal lunar tides, and this may be taken as the absolute minimum. The best procedure is to set up the tide gauge as early as possible, and to prolong the observations as long as possible. (Plate XIII facing p. 182.)

For the ordinary purposes of topographical mapping one station is enough. But if it is proposed to carry out precise levelling, and especially if there is any chance that questions of rising or tilt of the land will arise, then several tide stations will be required. In the present chapter we shall be content with one station.

It is essential that the tide gauge shall be set up on solid ground, so that there shall be no doubt of the permanence of the zero of the scale. It has usually happened that the longest series of tidal observations have been made at ports; but the value of these series has been much depreciated by the fact that ports are very often on made ground, with no guarantee against gradual settlement. Sometimes indeed it is found that the whole ground itself rises and falls with the tide.

The zero of the tide gauge must be connected with one point of the triangulation by a carefully observed line of levels.

A discussion of the instrumental details of tide gauges, or of the manner of reducing tidal observations, is entirely outside the scope of this book. Reference may be made to the *Publications* of the Survey of India, the *Handbooks* of this Survey, and to Sir George Darwin's work, *The Tides*.

The plane-table reconnaissance.

The success of the whole operation depends upon obtaining a good triangulation; and the difference between a good and a bad triangulation depends largely upon the efficiency of the reconnaissance which must precede the choice of the stations.

It is surprising how many small impediments combine to interrupt the mutual visibility of stations that might be expected to be in full view of one another. Hence it is not safe to take for granted the most seemingly obvious suitability of any particular station. A plane-table reconnaissance starts without taking much trouble over the measurement of a base, for the precise scale of the sketch is immaterial. Rays are drawn to three or four times as many points as are likely to be wanted, for it is certain that as the work proceeds it will be necessary to drop many of the points for obstruction on one ray or another. At the conclusion of the reconnaissance the surveying points will be carefully studied, from the point of view of planning a well-conditioned system of triangles, and of obtaining a good connection with the measured bases.

During the preliminary reconnaissance the ground should be beaconed, and care should be taken that not only the general localities are intervisible but the beacons themselves. Want of care in this respect may cause endless trouble afterwards, for the actual triangulation with the theodolite is a slow and laborious business, and the failure of a single station means that a number of other stations must be re-occupied. Hence no pains may be spared in making sure that every ray in the selected arrangement is really observable.

It is hardly necessary to say that the most anxious attention must be given to the selection of sites for the bases.

In suitable country the plane-table triangulation is easy. But it may be required to cross a forest-clad district where it will be necessary to build towers for the instrument, and the selection of the sites for these towers is difficult, because until they or their equivalents are erected it is impossible to judge of the suitability of their positions. To get over this difficulty one may, if transport is available, make good use of a reconnaissance ladder, such as that introduced into survey some twenty years ago by the then Commandant Durand of the French artillery. Such a ladder can be carried on a waggon and run up in an hour. From the opening in the top the plane-tabler can determine whether the position is suitable for a tower station. The same officer designed a very steady construction of poles that makes an excellent and inexpensive tower.

Plan of the triangulation.

It is not necessary that the principal triangulation should cover the whole country with a net of triangles. If the country is long and narrow a backbone of quadrilaterals is sufficient, as in the

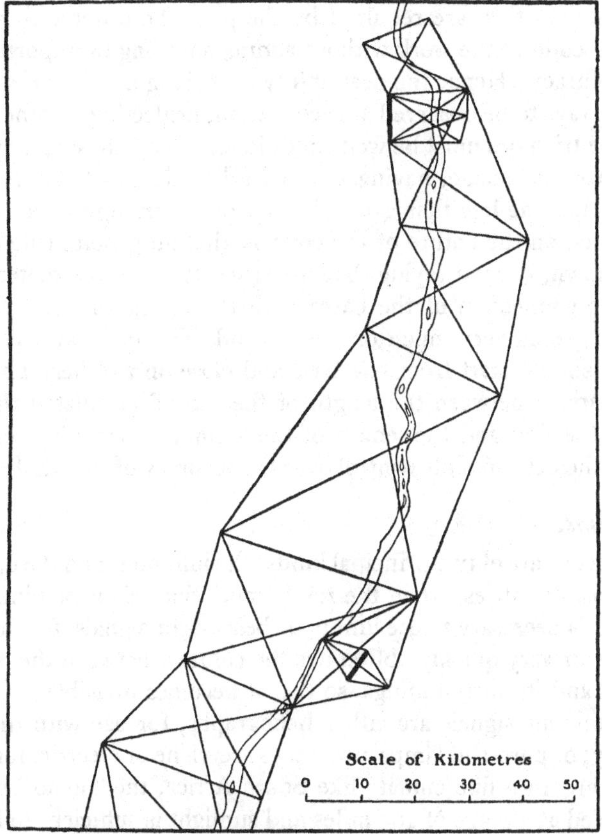

Fig. 9. *Diagram of Triangulation up the Nile Valley from Cairo to Beba. The two heavy lines are bases.*

figure. If the breadth is too great to make this sufficient, a chain of quadrilaterals at right angles to the first will strengthen the skeleton.

The purpose of forming the triangles into chains of quadrilaterals is to provide the necessary control without unnecessary repetition or duplication. Were the chain a chain of triangles only, there would be no check upon the survival of a gross error. A regular system of quadrilaterals gives the most simple arrangement consistent with strength.

It will often happen that more rays can be observed from a certain station than are required by the plan. To observe these rays would confuse the work without adding anything of importance to its accuracy. Hence the desirability of making a strict programme of the rays to be observed at each station, neglecting all others.

The triangles must be well conditioned: that is to say, they must be strong in shape, having no angles less than 30° about, and if possible none less than 40°. The size of the triangles will depend so much on the nature of the country that no general rule can be laid down, beyond saying that the larger they are the better.

The connection of the bases with the chains of quadrilaterals also depends very much on the ground. The main idea is that a chain should start from one base and close on another. Then the comparison between the length of the second calculated through from the first and the length of the second as actually measured gives the best possible control over the accuracy of the whole chain.

Beacons.

Beacons are of two principal kinds: the luminous and the opaque. In most countries, when the ray is more than eight or nine miles long it is necessary to use luminous beacons or signals, for the haze in the air very quickly obliterates the contrast between the opaque signal and its surroundings, so that it becomes invisible.

Luminous signals are either heliographs, for use with the Sun by day; or powerful lamps, nowadays acetylene or electric, for work at night. In a fine climate like South Africa, the "helio" can be observed at a range of 100 miles and sunlight is sufficient to let the work proceed without undue delay. In less favourable climates lamps provide a more certain mark for observation; but there may be difficulties in observing at night, for reasons of health or safety, that make it necessary to restrict operations to the day. On cloudy days it is sometimes possible to use powerful lamps instead of helios. Night observation has the great advantage that irregular refraction and disturbance of the air are much less than by day, so that the results are more accurate.

In either case, the effective use of luminous signals requires that the organisation and control of the signal parties shall be of a high order. Each party must use every endeavour to send its signal in the right direction so long as it is required, and must then move without delay to the next station. In general the discipline must be military to ensure the punctual carrying out of the lonely, dull, but all-important duties of the helio or lamp parties.

The United States Coast and Geodetic Survey have introduced automatic electric beacons run from accumulators, and turned off and on at the right times by clockwork, so that they can be left unattended like an automatic gas buoy for weeks at a time. In the open country in which so much of their work is done a small maintenance party with a car can control a large number of these beacons, especially if the stations are not too far apart. Modern American practice tends to use smaller triangles than formerly, as more economical in the end, because so much less time is wasted from poor visibility.

The heliograph.

It is impossible to enter into details of the mechanical construction of this instrument. Essentially it is a plane mirror from three to eight inches in diameter, mounted on a tripod in such a way that it may be turned by means of a slow motion screw to follow the Sun, and cast the reflected beam in a constant direction. There is a sight vane carried on an arm, which is set in the first instance upon the station to which the beam is to be directed; and a small round patch is left unsilvered in the centre of the mirror. So long as the black spot thus formed in the beam is kept centred on the white patch of the sight vane, the beam is being sent in the right direction.

It is sometimes a matter for surprise that the beam can be directed with the necessary accuracy, until one remembers that the Sun, and therefore the beam, has a diameter of half a degree. The light from the mirror therefore diverges in a cone of angle half a degree, equivalent to about 1 in 115. At a distance of ten miles the beam covers a front of about 150 yards, and a careful operator has no difficulty in directing the beam within this distance right or left of the object.

When lamps are used, at night, the distant station is invisible, and some means must be devised for sending the beam in the right direction. But it is hardly possible to go into such detail in this place.

It is interesting to remember that the limelight was invented by an officer of the Royal Engineers to make the connection between Wales and Ireland, after many weeks had been wasted in waiting for the Sun.

Opaque signals.

The opaque signals must be large, so that they may be visible at a distance; and they are best constructed so that the theodolite may be placed in position without taking down the signal; otherwise there is some danger that the signal will not be re-erected in its original position, and there will be a discontinuity between the observations made to that station before and after its occupation.

The construction of the beacons will vary with the materials locally available and with the ease or otherwise of transport. They should be symmetrical about their vertical axis when seen from any position; therefore the quadripod is preferable to the tripod. The appearance of a signal on active service is apt to differ greatly from its appearance on parade in the textbook.

In unsettled countries the surveyor has the advantage that he can cut down trees to make his signals; an official manual recommends a form of beacon built of 120 saplings. He can also clear the tops of hills of inconvenient timber, leaving the best tree standing in which to build a station. In civilised and closely settled countries this is not possible; but church towers and other buildings go far to supply the need. The Ordnance Survey of England constructed a station above the cross on the dome of Saint Paul's, and took off temporarily some feet of the top of the spire of Norwich Cathedral. Such operations being impracticable under ordinary circumstances, it is very difficult to carry out a piece of triangulation in England for instructional purposes on any considerable scale.

Permanent marks underground.

In all cases the visible signal should be considered as a temporary representation of an underground mark, which is buried for safety, and so protected and identified that it may be recovered at any time. To ensure this is not easy. A few of the principal stations of the Ordnance Survey are now lost, and a number of the stations of Struve's great Russo-Scandinavian arc of meridian, owing to insufficient attention to this necessity for permanent marking.

The form of the underground mark depends on local circumstances. It will take some such form as a copper bolt let into the rock at a depth of say two feet, and protected by a pyramid of masonry, or a large cairn of stones. In settled countries a few square yards of ground should be bought and enclosed; it may then be committed to the charge of the local authorities. In unsettled countries it is much more difficult to protect the stations, because the natural tendency of the native is to take the first opportunity of digging to see what it is that the white man has buried so carefully. In these countries also great trouble is often experienced in maintaining the opaque signals, and much time is wasted through their destruction during the course of the work.

Self-centering beacons.

With the usual construction of beacons a great part of the time taken in setting up the instrument is consumed in centering it over the station mark. And a considerable part of the gradual loss of accuracy in the progress of the chain of triangles is caused by errors in this setting up and centering. To avoid these difficulties an excellent form of self-centering beacon has been introduced by the Survey of Egypt. The beacon consists of a concrete pillar carrying on top a brass casting with three radial V-shaped grooves planed at 120° apart. The three levelling screws of the theodolite base, or the three feet of the helio, stand in these grooves. The instrument can be put up only in one position, which is automatically the correct one, and no error of centering is possible. When the station is not in use the pillar is covered over and protected in the usual way. This self-centering device seems to be worthy of general adoption.

Beaconing.

The selection of sites for the beacons is done by a reconnaissance party making a plane-table sketch well ahead of triangulation party, and erecting the beacons on the points which make the best conditioned system of triangles. The beaconing party is responsible for leaving everything in order for the triangulators; and they must take great care that all the rays are cleared and fully visible.

Initial latitude, longitude, and azimuth.

The triangulation and plane tabling will eventually produce a map in which all parts of the ground are shown in their correct relation one to another; but they will give no information as to the place of that country on the Earth. To fix the map down on the Earth, and to obtain its orientation, we must make astronomical observations.

By determining the latitude and longitude of one of the triangulation points we, so to speak, pin the map down at one point, leaving it free to turn about that point. By determining the azimuth of one of the rays from this point to any other, we fix the map in its right orientation.

Hence the necessary astronomical observations are

An initial latitude and longitude.

An initial azimuth.

It might be supposed that there would be some advantage in determining the latitudes and longitudes of a number of different points. But for our present purpose this is not so. The reason will be discussed more fully in the chapter on the astronomical observations for geodetic purposes. For the time being it will be sufficient to say that every observed latitude and longitude is affected by local irregularities in the direction of gravity, due to the unequal distribution of mass in the crust of the Earth. Hence if one observes the latitudes of two stations in the triangulation one will probably find that the difference of the observed latitudes does not correspond with the difference of latitude resulting from the triangulation. Such discordances are of the highest interest in Geodesy, and provide the material from which we derive a better knowledge of the Figure of the Earth and something of those peculiarities of its internal constitution implied in the theory of Isostasy. But in a simple topographical survey they are embarrassing; and it is usually best to confine oneself to a single latitude and longitude; and a single azimuth.

Observation of latitude.

The determination of latitude will be made by the observation of stars near the meridian: the method of circum-meridian altitudes. Observations of the stars are always to be preferred to observations of the Sun, both because they are more accurate in themselves, and because a number of stars can be observed on one night, so that a determination of the accuracy desired can be obtained very much more quickly than if the Sun were used.

Two or three nights' observation with a five-inch micrometer theodolite taking four pairs of north and south stars each night, should give the latitude correct to one or two seconds of arc, which is less than the probable value of the local deviation of gravity, and is sufficient for all topographical purposes.

Alternatively, the latitude will be determined, along with the local time, by observations of stars with the prismatic astrolabe. (See page 267.)

Should a more elaborate determination of the latitude be desired, it will be obtained with the zenith telescope, or with a large theodolite constructed to serve as a zenith telescope.

Observation of longitude.

The longitude of the initial station is the difference between its local time and the time of the meridian of Greenwich.

The local time at the initial station will be determined by the observation with the theodolite of pairs of stars east and west, as near the prime vertical as possible. Or it will have been determined along with the latitude by the prismatic astrolabe, or, more elaborately by a portable transit instrument set up at the initial station.

The Greenwich time, for comparison with the local time, will be received by wireless time signals. (See page 182.)

Observation of azimuth.

The azimuth is obtained by observation of stars east and west near the prime vertical, or of circumpolar stars at their greatest elongations. These observations give the difference of azimuth between the stars and the terrestrial station, and also the true azimuths of the stars at the moments of comparison; whence the true azimuths of the station are immediately derived.

If the triangulation is being made with opaque signals, so that no provision is made for illuminating the signals at night, it will generally be convenient to establish a supplementary mark at a moderate distance, say one mile, from the initial station. This mark can be more easily illuminated than one at a greater distance; and the difference of azimuth between the mark and the triangulation station chosen for the initial azimuth can be determined during the ordinary course of the triangulation by day.

There will be no difficulty in determining the initial azimuth in this way with an accuracy of two or three seconds of arc, which is sufficient for the purposes of the survey with which we are dealing at present. Local deviations of gravity have their effect on azimuths as well as on latitudes and longitudes; and it is useless to aim at a degree of accuracy which cannot be achieved except by the elaborate discussions of geodesy.

Azimuth cannot be determined with the prismatic astrolabe.

Wireless time signals.

The first regular service of time signals by wireless was established by General (then Commandant) Ferrié when he was commanding the post of military radio-telegraphy at the Eiffel Tower, about 1911. And to him is due in great measure the great technical development of the time service of the Bureau International de l'Heure (B.I.H.), officially controlled by a section of the International Astronomical Union and supported by subventions from the funds of that Union, but in reality owing almost everything to the liberal attitude towards it of the late Director of the Paris Observatory, M. Benjamin Baillaud, and to the enlightened policy of the French Government in allowing the use of its wireless stations for transmitting the signals.

The present system of the B.I.H. came into use on 1 January 1926, after discussion at the meeting of the Astronomical Union at Cambridge in 1925. (See *Geog. Jour.* LXVI, 242.) There are two types of signals, the International

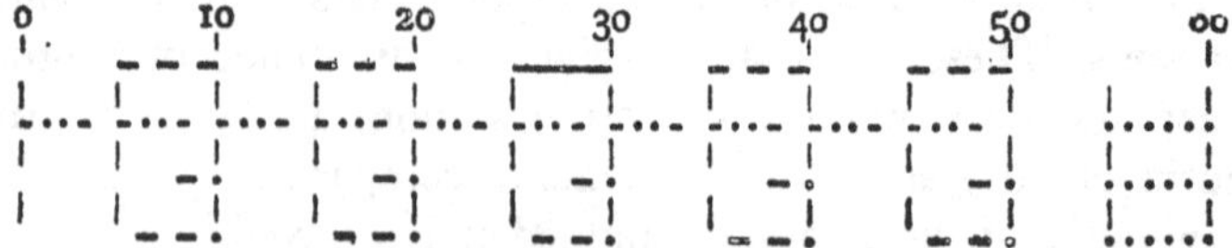

Fig. 10.

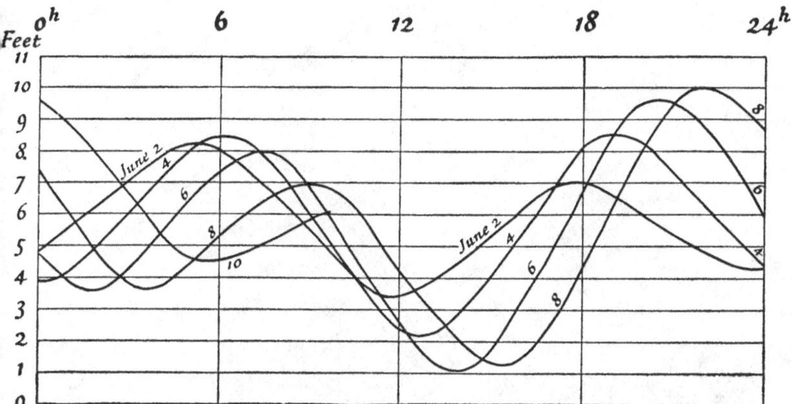

Reduced trace of the Tide-gauge Record for alternate days in the period June 2–10: the drum turns once in forty-eight hours, and the intervening days are on the other half of the sheet.

(*Geogr. Jour.* LXXIX, 203, *March* 1932

Portable Tide gauge of the U.S. Coast and Geodetic Survey pattern, as used by Mr Michael Spender at Low Isles, Great Barrier Reef.

(*Geogr. Jour.* LXXIII, 149, *February* 1929)

1. *Marconi R.P.* 11 *Long-wave receiver and frame aerial as used by Mr Francis Rodd in the Sahara. The observer is sitting on the aerial.*

(*Geogr. Jour.* LXXVIII, 524, *December* 1931)

2. *Phillips' Short-wave receiver permanently mounted on car, as used with mast-aerial by Major R. A. Bagnold in the Libyan Desert.*

and the Rhythmic

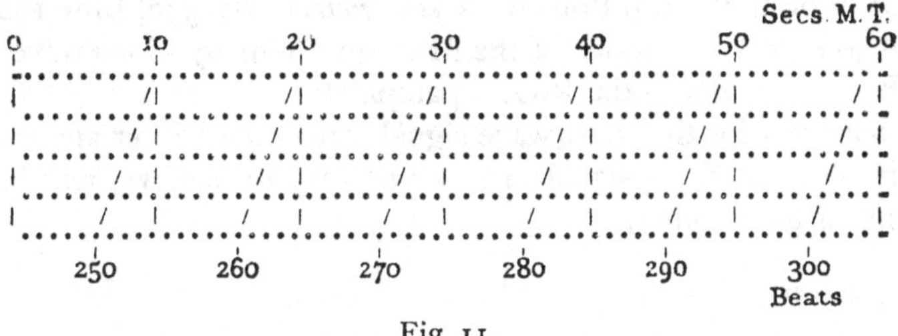

Fig. 11.

sent out twice daily at 08 and 20 h. G.M.T. by Issy-les-Moulineaux on 32 m. and by Bordeaux on 19,100 m.

The signals of the International type serve for navigation and any purpose in which an accuracy of one or two tenths in the comparison is sufficient. The Rhythmic signals allow very precise comparisons by the method of coincidences.

The German station of Nauen has for many years sent signals of similar types, recently made identical, at noon and midnight; and the American stations of Annapolis, Pearl Harbour, Balboa, and others, a series of dots at second intervals for five minutes twice a day.

Although all these signals were sent in Greenwich time, none of them came from the Royal Observatory until the establishment of the Greenwich time signal from Rugby in the autumn of 1927 at last removed the reproach that Great Britain was singularly unappreciative of the compliment the world had done her in adopting the meridian of Greenwich as the prime meridian for the time of the whole world. Rugby sends the Rhythmic signals only, at 10^h and 18^h.

The schedule of time signals, though now comparatively stable, is still liable to alteration and to extension, and it is useless to give details in a textbook. The Admiralty List of Wireless Signals gives the full details for every station in the world, with weekly supplements of corrections. It is for that reason somewhat cumbrous. The R.G.S. publish a pamphlet of the principal signals for the use of surveyors in their Technical Series.

Apparatus changes so fast that it is altogether useless to give any account of the construction and use of receiving sets in this book. A useful account of methods and instruments in 1925 is found in the description of the wireless work of Dr Hamilton Rice's expedition to the Rio Branco. (*Geog. Jour.* LXVII, 536, June 1926.)

A technical description of the Marconi Company's long wave set R.P. 11*a* is given in the R.G.S. pamphlet.

Receivers for the short wave signals are much lighter and more portable: but these signals are not so easy to observe, and have zones of inaudibility.

The bases.

Recent improvements in the means of measuring bases have altered our conception of the relation of the base to the triangulation. When base measurement was difficult one thought of the triangulation as built upon a single base, whose measurement was by far the most delicate, the most anxious, and the most uncertain part of the whole operation. If it were possible to measure a second base in another part of the country, well and good. But it would never have been thought remarkable if a whole survey rested upon one base.

Nowadays base measurement has become so relatively easy that it is possible to plan a chain of triangulation and to stipulate that bases shall be measured at intervals of 200 or even 100 miles apart all along the chain. The measurement of a base becomes a kind of control carried out at intervals as part of the routine; it is no longer the solemn and fundamental operation that it was.

Essentials of base measurement.

The essential steps in the process of base measurement are the following:

1. Determination of the field standard in terms of the national standard of length.

2. Measurement of the distance between the base terminals on the ground in terms of the field standard.

3. Correction of the measure for temperature and any other causes of variation, and reduction to national standard.

4. Reduction to the horizontal, and then to mean sea level.

For topographical purposes bases may be measured by means of tapes laid along a straight and level road or railway track, and stretched to a constant tension with a spring balance; or by a wire or tape hung under constant tension in the natural curve—the catenary—supported at each end by frictionless pulleys suspended from trestles, and carrying near each end a divided scale which is read against marks carried on tripods set up along the line of the base.

The tape laid on the ground serves very well when there is a convenient railway track with little traffic, such as is often available in large and new countries. In the absence of such ready-made sites for the base, it is easier to use the wires hung on tripods some feet above the ground. Much rougher country can be crossed, and it is easier to go up and down hill. We shall consider the latter method as the standard modern method, and shall deal with it first.

The tape or wire.

The modern tape or wire is either 100 feet or else 24 metres long between the divided scales. It is made of an alloy of nickel and steel (36 per cent. nickel) which has the remarkable property that its coefficient of expansion is only one-tenth that of ordinary steel, ·000,000,5 of its length per degree Fahrenheit, instead of ·000,006. The trade name of this alloy is Invar.

These tapes or wires have the great advantage that they are extremely portable. When wound upon their reels they may be sent by post to the National Physical Laboratory or to the *Bureau International des Poids et Mesures* at Breteuil, and there compared with the laboratory copies of the national standards of length. For a comparatively small fee the laboratory will determine the correction to the assumed length, and make a determination of the coefficient of expansion with temperature. Thus without any difficulty a survey department in, we will say, Fiji, may send home the standard tapes from time to time to be re-compared with the laboratory standards, and may thus ensure a very strict control over the lengths of the wires or tapes that are in actual use in the field.

This provides in a very simple and inexpensive way for the reference to the national standards of length.

Use of the suspended tape in the field.

The base will be from four to twelve miles long: the longer the better. It must be chosen so that one end is visible from the other; and it is generally convenient that the ends shall be on slight elevations, which much facilitates the choice of the stations for the base extension, that is to say, of the small triangulation connecting the base with one of the principal sides of the main triangulation.

The ends of the base will be marked by terminals sunk in the ground, or perhaps carried on concrete pillars.

Simple but firm tripods, carrying small upright pillars engraved with fine lines, will be set out in the line of the base, at distances apart equal to the tape length between its zeros of the engraved scales. Or in soft ground posts may be driven in and strutted.

The tape is suspended over frictionless pulleys, carried on straining trestles which can be adjusted easily so that the tape can be brought up close to the marks on the tripods. The tape is kept in tension by weights hung on each end.

At given signals observers at each end make series of simultaneous readings of the scale on the tape, against the marks on the tripod.

The tape is then carried on to the next tripod section, and the operation is repeated.

As soon as the tape party is clear a levelling party determines the difference of height between one tripod and the next. When this is done, the tripods in rear can be carried forward and set up ahead in the line of the base. With good organisation and drill it is possible to measure five kilometres per day in this manner.

At the beginning and end of each day's work the field tapes are compared with the tapes which have been standardised in the laboratory, and which are not subjected to the risks of injury in the field.

The accuracy obtainable in this way is limited only by the number of wires which are employed, the number of times that the measures are repeated, the precautions that are taken to avoid damage to the wires, and finally, by the residual error in the comparison of the standard tapes with the laboratory standards of length.

With very little precaution it is possible to measure topographical bases in this way with an accuracy of 1 in 200,000, which is amply sufficient for any work which does not aim at geodetic accuracy. We shall consider the refinements desirable in the measure of geodetic bases in a separate chapter.

1. *Taking readings on the wire, Semliki Base.*

Photos. by Capt. E. M. Jack, R.E.

2. *Adjusting the straining trestle and wire before reading.*

Plate XVI

1. *Levelling Staff of conventional pattern.*

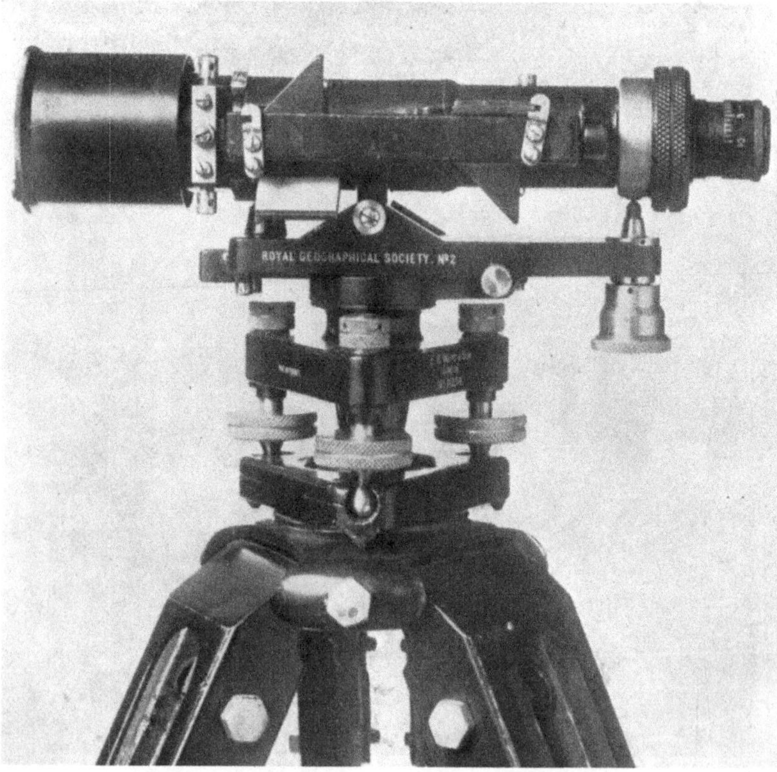

2. *Watts' Reversible Level, bubble read from eye-end by prism.*

Use of the flat tape in the field.

The tape laid on the ground will probably soon become obsolete, except for operations on a small scale.

A convenient way of operating with such a tape is as follows:

Provide the tape with two handles filed flat at their extremities. The distance between these extreme surfaces is the length of the tape which is compared with the standard. The handles are furnished with lugs, over which are hooked wire stirrups with a loop at the end, to take the hook of a spring balance.

Pickets are driven into the ground in the line of the base, at tape lengths apart, and strips of zinc are nailed to the tops of the pickets. The tape is held at one end by a spike through the loop of the stirrup, and is strained to the right tension by a spring balance hooked into the other stirrup. The flat ends of the handles should then come over the zinc strips. At a given signal observers at either end steady the tape, and draw a sharp point along the flat ends of the handles, so as to make marks on the zinc strips. The tape is then carried on to the next section, and the operation repeated. No attempt is made to join up the tape lengths accurately, but the distances between the pairs of cuts on the zinc are measured with scales or dividers, and the small excesses or defects added to the tape lengths. The thermometer is laid alongside the tape on the ground, and is read at frequent intervals to give the temperature correction.

Working in this way it is easily possible to measure a base with an accuracy of 1 in 75,000, provided that a good site can be secured. But unless a straight railway track is available the suspended tape is easier to manage though it requires a larger party.

For details of the use of tapes laid flat on the ground reference may be made to Wilson's *Topographical Surveying*.

Correction for slope.

With modern ideas of ground suitable for base measurement, the correction for slope sometimes becomes considerable; but it is always very simple.

The correction may be expressed in terms of either the vertical difference in height of the ends of the tape, or the slope of the straight line joining the ends.

Considered in its simplest possible form the problem is as follows:

Let L be the length of the tape, or the distance between the zeros at the two ends; and let H be the difference in height of the two ends. If θ be the slope of the tape, then $\sin \theta = H/L$; and the correction for slope is $-L(1 - \cos \theta) = -H^2/2L$ approximately for small slopes.

Hence, whether the slope or the difference of vertical height is measured, the correction to the measured length of each span is very simple. Modern practice favours the use of the differences of vertical height, measured with an ordinary **Y**-level, rather than slopes measured with an instrument such as the Abney level.

It may seem to be likely that the above simple assumption as to the form of the correction is not applicable to the case of a tape hanging in its natural curve. An exact investigation shows, however, that except in the most minutely refined work, the above formula is amply accurate.

Reduction to sea level.

It is an invariable principle that maps should be drawn as if the ground were projected on the sea-level surface. A moment's consideration shows that this is necessary. Think of two parallels of latitude crossing the Drakensberg from Natal to the Orange Free State. The ground on the latter side is nearly a mile further from the centre of the Earth than the low land in Natal. Hence the linear distance between the parallels will be greater by about one part in four thousand on the west side of the range. But it would be quite impossible to represent the difference on the map, since such a representation would require that the sheet should be stretched locally wherever the ground was high. Hence the necessity of reducing always to sea level.

The reduction is of great simplicity, being almost automatic in its operation. Let h be the mean height of the base above sea level, and R the radius of the Earth, supposed spherical. Then the simple reduction of the measured length L of the base is evidently $-Lh/R$ very nearly. This is quite sufficient except in extremely accurate geodetic work; the latter we will consider in a later chapter.

With our base thus reduced to sea level the calculation of the whole triangulation, depending on the base, is automatically reduced to sea level also, as we shall see almost immediately.

The theodolite triangulation.

The theodolite is set up directly over the station mark, and is levelled, so that the horizontal circle of the instrument is truly horizontal. The telescope is sighted on each of the other stations in turn, according to the programme; and the horizontal circle is read by the microscopes at each setting. It is important to notice that the angles thus measured are not the actual angles between the successive pairs of beacons, such as would be measured by a

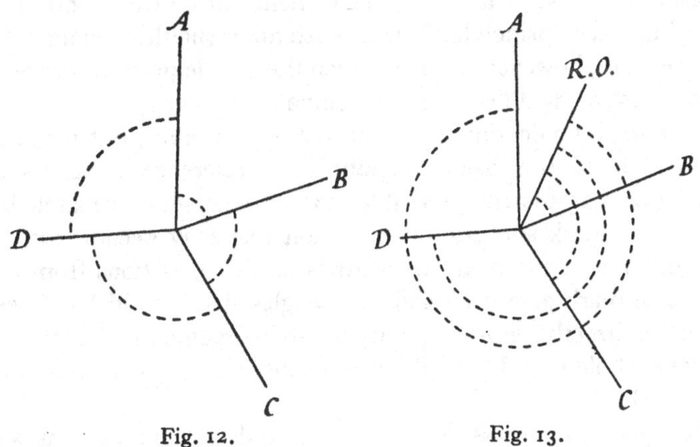

Fig. 12. Fig. 13.

sextant, but are those angles projected on to the horizontal plane of the station occupied. If the Earth is considered spherical these angles are the same whatever the height of the stations above sea level, even when the various stations are at very different heights. Hence the justification for the statement of the preceding section, that when the measured base is reduced to sea level the whole of the triangulation is automatically reduced to sea level also.

The details of the manipulation of the theodolite are dealt with in the *Textbook of Topographical Surveying*. We will confine ourselves here to the points which are especially concerned with the horizontal triangulation.

Suppose that there are four stations A, B, C, and D to be observed. A round of angles consists of settings on A, B, C, D, and

A in turn. It is important to notice that the station *A* is observed to twice. The reason for this is that small errors are introduced into the measured angles if the instrument is not perfectly stable during the course of the round; and this cannot be ensured absolutely. If we repeat the observation of *A* at the end of the round we ensure that the measure of the angle *DOA* is independent of any settlement of the instrument that may have occurred during the course of the round from *A* to *D*. In fact any such movement affects only the particular angle under measurement at the time when it occurred. At the same time the near concordance that there should be between the first and the last settings on *A* is a measure of the stability of the instrument during the round. This is an important principle, to which attention should be paid.

There are, however, cases in which the simple process described above must be modified. In a bad climate it will happen that sometimes one and sometimes another station is visible; but never all at once. In such a case one establishes a reference mark at such a distance that it is always visible; and one measures the angle between this mark and any other station that may become visible. The differences between the bearings of all the stations from this reference mark give eventually the angles that should have been observed directly. For simplicity we shall assume in what follows that the whole round of angles can be observed in the manner first described.

To obtain the necessary control and to detect blunders, as well as to eliminate by repetition the accidental errors of the observations, the round of angles must be repeated one or more times. We take care to arrange the work so that each repetition is made on a different part of the circle, whereby accidental errors of the circle graduation are to some extent eliminated. In the very best instruments these errors rarely exceed one second of arc; but there are many instruments in use, made by well-known makers, in which the errors of graduation are by no means so small. It would be a very long and tedious business to determine them and introduce corrections for them; and it is usually sufficient to so arrange the observations that repetition tends to eliminate them.

Check on the accuracy of the triangulation. Triangular error.

In a small triangle, of one or two miles per side, the curvature of the Earth may be considered negligible, and the three angles of the observed triangle should add up to 180°. In larger triangles this is not strictly the case. We remember that the angles are measured in the horizontal planes through each station; and when the triangle is large the curvature of the Earth throws these planes out of parallelism with one another. The result of this is that the three angles of the triangle should add up to more than 180°, the excess being called the "spherical excess". This is easily calculated, and depends on the area of the triangle. Deduct it from the observed sum of the three angles, and the result should be 180° exactly. In practice it will differ from this quantity by a small number of seconds, and this difference is called the "triangular error".

The average triangular error is a measure of the precision of the triangulation. In the most precise work, of geodetic accuracy, the average error will be less than one second of arc. In good topographical triangulation it will be two or three seconds. Experience will show what limit should be set to the average triangular error on any particular piece of work; and when the standard is laid down the observations must be repeated as many times as may be necessary to arrive at the desired degree of accuracy.

This test of the triangulation by considering the triangular error gives a simple and invaluable means of control, and of securing that the work is up to the standard required.

Spherical excess.

The spherical excess of a triangle may be calculated from the formula

$$E = ab \sin C \frac{\operatorname{cosec} 1''}{2\,(\text{radius})^2}.$$

When the ellipticity of the Earth is taken into account it is usual to work with the radius of an oblique section making an angle 45° with the meridian. See Auxiliary Tables of the Survey of India, Table III.

If the triangles are less than 100 miles a side it is sufficiently accurate to apply to each angle a correction of one-third of the excess.

Except in very accurate work it is sufficient to measure the area of the triangle from a chart, and to use the formula

$$E = \frac{\text{area of triangle in sq. miles}}{1000} \times 13''\cdot 15.$$

Refraction of horizontal angles.

Refraction is usually supposed to act in a vertical plane; and so long as it does so it can have no sensible effect upon the horizontal angles. But in the neighbourhood of steep slopes the layers of air of different temperatures are not horizontal, and a ray of light passing through such a region may suffer deviation in the horizontal plane. Such effects will in general take place only near the ground; and for this reason it is important to avoid rays which in their course approach near intervening ground. Such rays are called "grazing rays".

It is difficult to say how high a ray must pass over intervening ground to avoid the disadvantage inherent in a grazing ray; but so long as obviously grazing rays are avoided it is usually safe to hope that such small remaining effects of horizontal refraction as may be left will be of a non-systematic character, and will be eliminated in the average. There are, however, certain cases where this is not necessarily true, such as in the triangulation up the valley of the Nile, where the stations are alternately on opposite cliffs, and the cooling of the air over the water might very well produce systematic effects.

A publication of the United States Coast and Geodetic Survey, on the Texas-California arc of primary triangulation, gives a striking instance of well-determined lateral refraction. The line between two stations Clayton and Kennard passes very close to a steep slope of a flat-topped hill. During most of the observations the wind was blowing from the hill across the line between the stations, and the results gave excessive closing errors to the triangles involving this line. The observations made when the wind was blowing across the line towards the hill gave values which closed the line in a satisfactory manner. It became evident that the former were in error by about seven seconds of arc.

Calculation of the triangulation.

The observations made at each station are entered in "angle books", from which the observed angles are abstracted, the means taken, and the results brought together in the form of "abstracts of angles".

Let the figure represent a part of a chain of quadrilaterals under observation. A portion of the abstract of angles may read thus:

Observed angles at Y:				
	EYD	31°	57′	28″
	DYF	45	6	15
	FYE	282	56	9
		359	59	52

It will be noticed that the angles at the station do not add up to precisely

360° as they should. This is due partly to the errors of observation and partly to movement of the instrument in the course of the round.

Now suppose that in the course of the computation we have arrived at the length of the side *DE* and that we wish to proceed. Let us do so by solving the triangle *DEY*. From the abstract of angles at *Y* we take out

EYD 31° 57′ 28″

and from the abstracts of angles at *D* and *E* we take out

	YDE	55°	28′	16″
	DEY	92	34	42
The sum of these is		180	0	26

and by calculation we find that the spherical excess is inappreciable, which leaves the triangular error

+ 26″

Now we cannot solve the triangle until it has been made into a perfect triangle, whose angles add up to exactly 180°. If we were concerned with this triangle alone, the best thing that we could do would be to divide up the error into three equal parts and apply one to each of the angles, so that the sum of the thus corrected angles is exactly 180°.

This being done we could find the length of either of the sides *EY* or *DY*, being given the other side *DE* and the angles of the triangle.

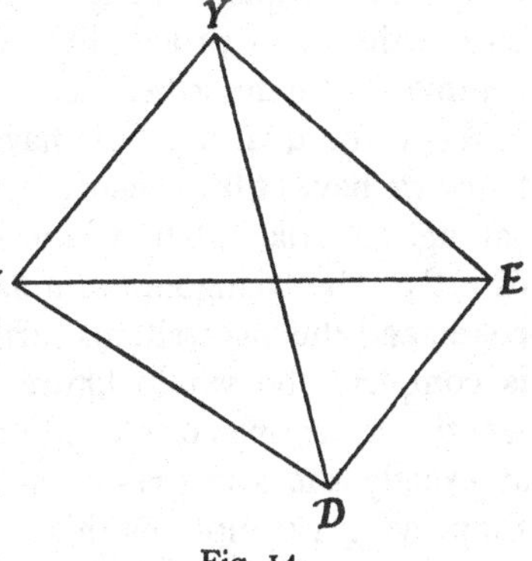

Fig. 14.

For by elementary trigonometry

$$\frac{DY}{DE}=\frac{\sin DEY}{\sin DYE},$$

whence $DY = DE \operatorname{cosec} DYE \sin DEY,$

and similarly $EY = DE \operatorname{cosec} DYE \sin YDE.$

Thus if $\log DE = 3{\cdot}8337549,$

$\log DY = \log DE$	3·8337549
$+ \log \operatorname{cosec} DYE$	0·2763299
$+ \log \sin DEY$	$\bar{1}$·9995610
	4·1096458

and similarly $\log EY = 4{\cdot}0259149.$

But now suppose that we proceeded to find the length of *FY*. We might do it from the triangle *FYE*, since *YE* has been found; or from *FYD*, since *YD* has been found. But in solving either of these triangles we should be confronted with the same difficulty, that the sums of the three angles were not precisely 180°. If we made them so in the arbitrary way supposed used in the first triangle, we could make the solutions, but should arrive at slightly different values of *FY*. And if we proceeded to find the length of *FY* by way of *FD* and *FE* we should arrive at two other values.

In very rough work we might be content to take the mean of the results and proceed; but this will not do in accurate work. We must not apply to the angles *FDY* and *DYE* arbitrary corrections which are inconsistent with an equally arbitrary correction to their sum *FDE*. Evidently we must find a set of corrections to the individual angles in the quadrilateral *FYED* which shall simultaneously reduce the sums of the three angles in every triangle to 180°. This is called the adjustment of the quadrilateral. The method is based on the theory of probability, and even in this simple case is too complex to be described here.

When the quadrilaterals have been reduced to internal consistency we have still to ensure that the length of one base calculated through the triangulation from another is the same as that measured directly. The triangulation must be made to "close" upon all the bases, and this necessitates further adjustment. When the process is complete the whole figure is consistent. One may calculate lengths or bearings or co-ordinates round by any path, and arrive at exactly the same result as by any other path, but even with simplifying assumptions this is a complex process.

Calculation of the geographical positions of the triangulation points.

For general purposes the most convenient way of defining the position of each point is to give its latitude and longitude.

We have determined an initial latitude and longitude of a station *A*, the azimuth of the line *AB*, and the length in feet of *AB*. The problem is to determine the latitude and longitude of *B*, and the azimuth of *A* at *B*. For the purposes of precise survey it is not sufficient to resolve *AB* along and at right angles to the meridian through *A*, and convert these resolved distances into differences of latitude and longitude. We have to take account of the curvature of the Earth, and of the fact that owing to its spheroidal shape the curvatures along and at right angles to the meridian are not the same. The necessary formulae are somewhat complicated, and the analysis by which they are established still more so. It must be sufficient to say that with suitable auxiliary tables the process is straightforward and not difficult. The elegant method due to Puissant and adopted by the Survey of India has in the last edition of the *Textbook of Topographical Surveying* been abandoned in favour of the method due to Clarke, less elegant, but presumably more accurate.

We should note that this process demands a knowledge of the size and shape of the Earth, as a basis for the tables.

The calculation gives the differences of latitude and longitude between *A* and *B*; which being added to the co-ordinates of *A* give those of *B*. It gives also the azimuth of *A* from *B*, technically called the "reverse azimuth", which is not quite 180° different from the initial azimuth, on account of the convergence of the meridians towards the pole.

Now to proceed from *B* to *C*: We have the measured angle *ABC* and the azimuth of *BA*, from which we deduce the azimuth of *BC*, and the process of carrying on from *B* to *C* proceeds as before.

Thus we obtain the geographical positions of all the triangulation stations. Any selection of them can now be plotted on any plane-table sheet in the manner to be described presently.

Calculation of rectangular co-ordinates.

If however the area surveyed is small, and it is a question only of plotting a single sheet or small number, it may not be worth while to perform the above calculation of the latitudes and longitudes. For the purpose of plotting the plane-table sheets it is sufficient to calculate the rectangular co-ordinates of the other stations with respect to axes through the initial station, treating the triangulation as if it were on a plane instead of being on the spheroid. This is simpler to calculate, but we shall see that it is not so simple to plot; and there is great difficulty eventually if geographical positions should be wanted after all.

Determination of heights by theodolite, or trigonometrical heights.

The method of trigonometrical heights provides a rapid way of determining differences of height of the triangulation points.

During the occupation of each station, after the horizontal angles have been measured, the apparent angular elevation of the beacons at the other stations are measured with the vertical circle and microscopes of the theodolite. These measures differ from the horizontal angles in that they are absolute elevations or depressions, and not merely differential measures.

By careful attention to the rules for the measurement of vertical angles the errors of the instrument may be very nearly eliminated, except that there is no possibility of eliminating the errors of division of the circle by repetition on different parts of the arc. The serious errors inherent in this method are those due to refraction.

The vertical refraction of a horizontal ray is large; it depends upon the density of the air intervening in the path of the ray, which varies very rapidly during the day, and is almost impossible to calculate. But there are two principles of working by which this effect may be in great part avoided: the observations should be made at the time of day when the refraction is a minimum, that is to say, in the early afternoon; and the rays should be observed from opposite ends in circumstances as nearly the same as possible. It is then assumed that the effect of refraction on the observation at each end is the same; and it is easy to see that the effect is thereby eliminated.

When it is not possible to take the reciprocal observations from each end, some assumption must be made as to the law of refraction, based upon an analysis of the observations that have been made reciprocally. But to pursue this part of the subject is beyond the scope of the present chapter. The results are especially unsatisfactory in mountainous regions with glaciers and perpetual snow. Hence the heights of the inaccessible snow-clad peaks of the Himalayas, which are determined by this method, are subject to some uncertainty.

It will be noticed that when two points at a considerable distance apart, and of no great difference of height above sea, are reciprocally observed, each is measured as a depression at the other, in spite of the fact that refraction raises each of the rays to some extent. This is a necessary consequence of the curvature of the Earth; but the size of the effect is a little surprising to the student, until he remembers how small the Earth really is, and that two points four geographical miles apart subtend an angle of four minutes of arc at the centre of the Earth, so that each is depressed from the other by two minutes.

The formulae are very simple.

Let A, B be the two stations, and M a point vertically below B at the same distance from the centre of the Earth as A.

Since AB is very small compared with AC, we may take

$$BM = AM \tan BAM \qquad \text{(Fig. 16).}$$

And it is easily seen that BAM is equal to half the difference in the *depressions* of the two stations, seen from one another: provided that the refraction is the same at the two ends.

Hence if s be the distance in feet between the two stations, their difference of height in feet is $s \tan \frac{1}{2}$ (difference of depressions).

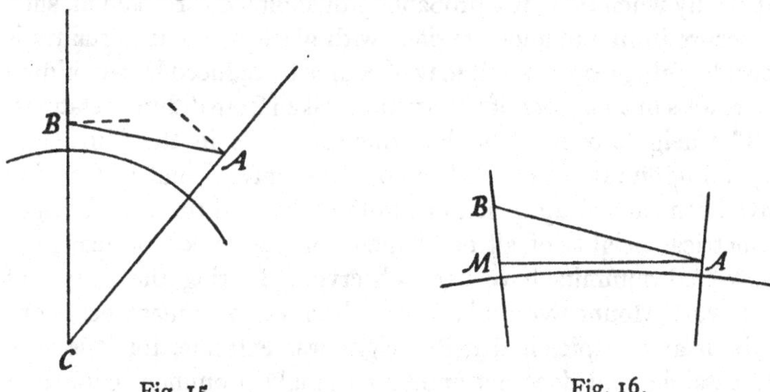

Fig. 15. Fig. 16.

If only one angle is observed it is necessary to introduce a numerical allowance for refraction. Whenever reciprocal observations are obtained, as above, the refraction at either station is $R = \frac{1}{2}$ (θ – sum of observed depressions) where θ is the angle ACB, which can be calculated from s and the radius of the Earth. It is found that the ratio $R : \theta$ is generally about 0·07.

Now it is easily seen that for an observed depression D at B the angle

$$BAM = \tfrac{1}{2}\theta - \text{refraction} - D.$$

And if refraction $= 0{\cdot}07\theta$ this becomes

$$BAM = 0{\cdot}43\theta - D.$$

Or since the distance along the surface corresponding to an angle of 1″ at the centre is about 101 feet

$$BAM = \frac{4''{\cdot}25 \times \text{distance in feet}}{1000} - D$$

$$= \kappa - D \text{ say.}$$

And then as before

$$\text{difference of height} = s \tan (\kappa - D).$$

This is the simple theory on which the table for κ given in the *Textbook of Topographical Surveying* is based.

The whole of the above theory is a rough approximation; but it is probably sufficient, in view of the uncertain effects of refraction.

We must also take into account in the calculation the height of the theodolite above ground at each station, and also the heights of the signals to which the observations are made. It is tedious to puzzle out the rules of signs for these small corrections; but it is quite easy to apply them when the rules are given. See *Textbook of Topographical Surveying*, pp. 50 et seq.

When all precautions are taken the errors of this process amount to one or two feet on a ray of thirty miles, when the angles are observed at each end. For rays observed only from one end, and especially when there is a probability of anomalous refraction, such as occurs in mountainous regions with glaciers, the uncertainty is considerably greater; but it may of course be reduced by combining the results of a number of observations taken from different stations.

The heights of most of the principal peaks in the Himalayas depend upon rays observed from one end only. Lines of levelling have been carried up to stations in the hills, and from these trigonometrical heights of all or of many of the 10,000 permanently snow-clad summits have been observed. During the course of such work Mount Everest had been observed to on many occasions without any suspicion that its height was extreme, for it is very inaccessible, and does not appear strikingly prominent compared with other peaks. Only when the observations were computed in due course in the Survey Department at Calcutta was it found that this peak is higher than any other known.

Construction of the plane-table graticule.

The plane-table graticule is a construction of meridians and parallels drawn on the table, so as to permit the plotting of any station whose latitude and longitude have been derived from the preceding calculation. The projection upon which the graticule is drawn is an approximation to the polyconic projection. The quantities required are derived from tables such as those given in the *Textbook of Topographical Surveying*, pages 298 et seqq., and the method of construction is given on page 105. See also the author's *Map Projections*, pages 58, 118. This work can be done in the field, with only the sight rule of the plane table and a pair of dividers. When the graticule is plotted it is easy to interpolate and plot any station whose geographical position has been determined.

At first sight this method may seem to be elaborate, and the beginner may imagine that something much simpler might be devised to serve its purpose. It will be found, however, that for the orderly conduct of a regular survey nothing less systematic will serve. Generally speaking, when the triangulation is complete and calculated, the work of plane tabling the detail is divided up between a number of surveyors, who undertake each a definite block bounded by certain meridians and parallels. Each plane table will be plotted so that it includes an area somewhat larger than that assigned to the particular table; and the triangulation points will be plotted all over the sheet. To have these well-determined points all round outside the area actually being mapped is a great convenience.

Mapping the detail by plane table.

When the surveyor goes out into the field with his plotted sheet he first visits some of the principal points, sets up the table there, and sights round to the other points within view, to see that the rays come right, and that no mistake has been made in the plotting. At one of the stations, while the table is set, he puts on the compass lines for the trough compass. He also takes the opportunity of drawing rays to prominent objects which will serve as intersected points.

This being done, the work of mapping proceeds as in graphical plane tabling, and we need not repeat what has been said on the methods of fixing by resection. Since the positions of the calculated points are correct far within the limits of visible error, the general accuracy of the whole will be much better than in the graphical process, and a second resection should always give an exact cut of the three rays, without any triangle of error.

The determination of heights and contours will be very much more exact than in the graphical process, because they will be based on the theodolite heights of the triangulation stations.

The spot heights at the resected points will be determined with the Indian pattern clinometer; and being based upon accurate fundamental heights, it gives points for contouring with all the refinement which can be desired, except for detailed engineering.

When the spot height is fixed, the slopes are observed with the scale of degrees on the other side of the opening in the vane, and the processes of calculating mentally the places and spacing of the contours are almost exactly similar to those which we have already described fully in Chapter VIII, pages 147–151.

When the section allotted to the surveyor is complete he inks in all the lines and figures which are to stand, clears up with rubber all those which are not required, and sends in the field sheet to headquarters. The great advantage of this system is easily seen. The field work comes in block by block all ready for the draughtsman, who has only to redraw it on the finished sheet, when it is ready for reproduction. There is very little possibility of confusion or misunderstanding, or of some portion of the work being missing. All this tends strongly to good order, efficiency, and economy in the operations of the survey.

Precise traversing.

It must always be remembered that in certain types of country triangulation and plane tabling are impossible, as on the Gold Coast, a nearly flat and densely forested region, where it is impossible to obtain a view of any importance. In such country it is necessary to construct a framework of traverses, run with the theodolite and steel or invar tape along lines of clearance cut through the forest. This is expensive, and exceedingly unsatisfactory, because in a few months the lanes are grown up again, and the labour seems to be lost.

The methods of work vary much in different countries, and for an account of them we may refer to such books as the *Handbook of the Southern Nigeria Survey*.

Tacheometer and subtense traverses.

The principle of subtense measurement is simple. It is required to measure the length of a ray over rough country unsuitable for ordinary traversing. At one end of the ray two marks are erected in a line at right angles to the ray. Their linear distance apart is measured by the party who erect them; and their angular distance apart is measured with a theodolite from the other station.

Then if 2θ is the angle subtended by a length $2s$, the distance is $s \cot \theta$.

This is the method generally employed on long rays. The marks are poles set up perhaps fifty feet apart, and the angles are measured by the method of repetition. The method has often been used on boundary surveys.

Over shorter rays it is possible to invert the process, and instead of measuring the angle which is subtended by a definite distance, one measures the distance included in a fixed angle. Wires or marks at fixed distances are inserted in the field of the theodolite or level, and a graduated staff like a levelling staff is observed. The further away the staff, the greater is the length of it included in a fixed angle of sight. The fixed marks in the telescope, called stadia marks, are standardised so that one has the factor, frequently 100, by which one multiplies the length read on the staff to obtain the distance of the staff from the observer.

The factor evidently varies with the distance of the stadia marks from the optical centre of the objective of the telescope, which is changed in bringing to focus objects at various distances. It is not hard to prove that the necessary correction for this can be obtained by the simple process of adding, to the distance computed as above, the distance from the centre of motion of the instrument to a point on the axis, outside the objective, at a distance from the optical centre equal to the focal length. Thus there is a small and slightly variable correction to be added to the observed distance to obtain the correct result.

This is not difficult to do; but the necessity for doing it can be avoided by introducing into the telescope a third lens, called the anallatic lens, which eliminates the small correction just described, and gives at once the true distance from the centre of motion of the instrument. A theodolite or level fitted with this device is called a tacheometer. The instrument is more useful in making detailed plans of a small area than in geographical work.

Correction for slope.

There is often some confusion as to the correction required when the ray is observed on a slope. Without going into the proofs, which are quite simple, we may state the rules shortly as follows:

When the subtended angle is measured on the horizontal circle of the theodolite, no correction for slope is required.

When stadia lines or tacheometer are used with a graduated bar placed horizontal, the apparent distance must be multiplied by the cosine of the slope to give the horizontal distance.

When the graduated bar is placed vertical, the apparent distance must be multiplied by the square of the cosine of the slope.

The processes of topographical survey which we have described here are those which, with small differences, are in use in all the principal surveys of the world. They are admirably suited to the rapid production of maps on topographical scales, especially where the detail is not too crowded.

They are not, on the other hand, so suited to the production of large scale cadastral maps showing intricate boundaries of property. In such cases it is usually necessary to break down the triangulation into small triangles of about a mile a side, by the theodolite, and to cut these up in turn and fill in the detail with the chain. (See page 212.)

Levelling.

Trigonometrical heights are good enough for a topographical map of a new country; but the time will come when there is a demand for more accurate and more detailed knowledge of the relative heights, and the survey cannot be considered complete until lines of levels have been run all over the lower and more accessible and developable parts of the country. These lines will be run, normally along the roads, with a level and pair of staves.

Putting aside for the time all the details of construction and adjustment of the instrument, the level may be described as a telescope which is set at right angles to a vertical axis, and can be swept round in a horizontal plane. Any object seen on the cross wires of the telescope is on the same level with them, or with the centre of the object glass of the telescope.

The levelling staff is a pole, usually rectangular in section, with a face divided into feet, tenths, and hundredths of a foot, numbered from the bottom upwards (Plate XVI facing p. 187).

Suppose that this staff is stood upon the ground at a distance from the level, and that the observer, looking through the telescope, sees that the horizontal wire cuts the image of the staff at the height 7·61 feet, whereas the centre of the object glass is 4·27 feet above ground. It is clear that the ground at the base of the staff is 7·61 − 4·27 = 3·34 feet *below* the ground on which the level stands, provided that the telescope is really pointing horizontal.

Similarly, if the reading of the horizontal wire is 9·03 feet upon a second staff, the ground at the foot of the second staff is 9·03 − 7·61 = 1·42 feet lower than at the foot of the first. And it should be noticed that this result is obtained without requiring any knowledge of the height of the object glass above ground, which varies each time the instrument is set up, by variation in the spread of the legs of the tripod, and is not quite conveniently measurable in ordinary instruments. The essence of the method is the use of the two staves, so that the height of the level itself is eliminated.

In the above explanation of the use of the level we have made the important reservation "provided that the telescope is really pointing horizontal". The surveyor in the field cannot be continually revising the adjustment of his instrument, and it is characteristic of a good method of survey that the way in which the instrument is used should eliminate automatically errors due to imperfect adjustment.

In the use of the level this end can be attained with great simplicity. All that is necessary is that the two staves should be placed at equal distances from the telescope. The reason for this will be seen at once, when we consider in what way the instrument is likely to get out of adjustment. Reduced to its simplest parts, the level consists of a telescope fixed on an axis *V*, with a bubble *B* firmly fastened to the tube of the telescope. The axis *V* must be set up vertical, and the test of verticality is that the ends of the bubble *B* should not move along the scale when the instrument is rotated about *V*. The bubble may not be central on the scale, but so long as it does not move along the scale when the instrument is rotated, the axis is vertical. Hence the process of getting the axis vertical, which is required every time the instrument is set up, does not in theory depend upon the adjustment of the bubble or of the telescope.

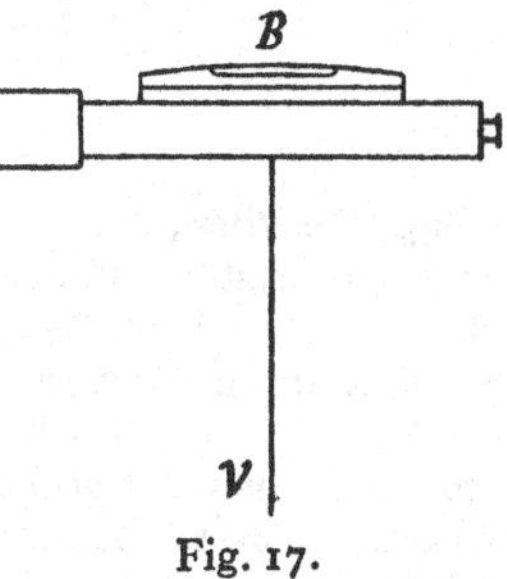

Fig. 17.

Now if the bubble is not in the centre of its run when the axis is thus found to be vertical, the bubble is out of adjustment. And even if the bubble is in adjustment, it does not follow that the telescope below is parallel to it.

These errors are adjusted from time to time by the methods given in any textbook on the use of the instrument. But they are apt to vary from day to day or from hour to hour while the instrument is in the field. The former does not affect the results if care is taken to make the test of verticality of the axis as above, not that the bubble is central, but that it remains at rest when the instrument is turned about the vertical axis. But the effect of the second error is that the telescope is pointing up or down at some small angle to the horizontal when the axis of rotation is made truly vertical. The resulting error of reading on the staff will be strictly proportional to the distance of the staff. Hence if two staves are at the same distance, the errors in reading will be the same, and the difference of the two readings, which is the quantity required, will be unaffected by the error.

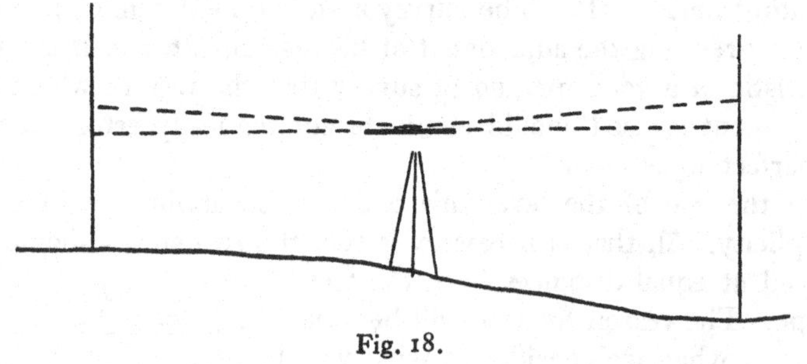

Fig. 18.

Suppose, then, that we wish to run a line of levels from *A* to *B*. We mark out suitable stations *K*, *L*, *M*, ... in the line *AB*, for the staves. We set up the level midway between *A* and *K*, and find the difference of height of the ground at *A* and *K*. Then we set up midway between *K* and *L*, and so on. Each difference of height is then independent of the errors of adjustment of the instrument, and it is not possible for the effects of these errors to accumulate, since they are automatically eliminated at each step.

We should notice also that this process eliminates the effect of the curvature of the Earth. The line of sight of the telescope, being perpendicular to the vertical of the observer, is a tangent to the sphere and consequently passes higher and higher above the surface as it proceeds. Owing, then, to the curvature of the Earth, all staff readings are too high by a very small amount, which would become serious in time if sights were taken always forward from one station to the next, but is eliminated without conscious trouble by the method of setting up the instrument always half-way between the two staves. Thus no correction for the curvature of the Earth appears in ordinary levelling operations at all, even though they extend over hundreds of miles. Paradoxers who imagine that the Earth is flat are fond of quoting this as a confirmation of their idea.

In a precisely similar way, there is no need to take any account of refraction if the level is always set up midway between the two staves. The effect of refraction will be to raise each staff in the field of view of the level, so that the readings are slightly too small. If observations were taken always to a forward staff this effect, extremely small for any one ray, would accumulate steadily, and would be very difficult to calculate. It is important, therefore, to notice that by the proper use of the two staves the effect of refraction is completely and automatically eliminated.

In work of the highest refinement it is however not quite good enough to say that the curvature of the Earth and the effects of refraction are entirely eliminated by the simple process of setting up the level exactly half-way between the two staves. Since the Earth is not strictly spherical, but spheroidal, the curvature varies continuously along a line of levels, unless it runs along a parallel of latitude; and in a very long line there will be a very small but cumulative effect, which must be taken into account. Similarly, if a long line of levels is taken over country continually rising, the forward ray is always nearer to the ground than the back ray, and passing through air of higher temperature if the sun is shining. Again, as the refraction decreases with increasing temperature until about three o'clock in the afternoon, the forward rays, which are observed after the back rays, will be taken systematically in conditions of less refraction until the middle of the afternoon, when the condition is reversed. These slight systematic refractions can be in great part eliminated by arranging that at the repetition of the work the effects shall be reversed in sign.

We should note also that since the bubble sets itself at right angles to the direction of gravity, which is not necessarily normal to the assumed "figure of the earth", owing to local deviations of the vertical arising from irregular distribution of mass, the level surface is somewhat irregular in shape. The deviations have no practical importance, because of course the level surface determined by instrumental levelling is the same as that of a free surface of water. The difference between the spheroid and the "geoid" is however of importance in geodesy, and will be referred to again in Chapter XI.

Bench marks.

The familiar horizontal line and broad arrow, seen so often along the English roads on walls and gate-posts, or the broad arrow without the line on kerb stones, are the bench marks left by the levellers; and the corresponding heights above sea are all marked upon the six-inch map, to a tenth of a foot. These are secondary bench marks, liable to displacement by sinking, reconstruction of walls, and similar accidents, and many are now hard to find. A "bench-mark walk" with a six-inch map is an instructive amusement for schools. The spot heights of the six-inch map, mostly on the crown of the road, are given only to the nearest foot, and are not bench marks.

Detailed sections.

If we wish to make a section in detail for drainage schemes we may modify the above process without any damage to the essentials.

The staff may be set up at any number of intermediate points between *A* and *K*, and the difference of height from *A* or *K* found. These observations will be affected by the instrumental error, but since the distances are short the effect will be small. And these observations will be used only to refer points between *A* and *K* to either *A* or *K*; nothing will be carried on from them into the next section, which will start from *K* unaffected, as we have seen, by the instrumental error.

Survey for artillery.

The special methods of survey for artillery played so important a part in the Great War that the following brief account of them, as written for the second edition of this book, may be of interest, if only historically. No attempt has been made to bring the account up to date. The Field Survey Battalion provides bearing-pickets well distributed over the gun positions, with lists of their coordinates and of the bearings of several well-defined points from each picket. The battery has an Artillery Director, which is essentially the telescope and horizontal circle of a theodolite, with an erecting eyepiece on the telescope. Each gun is provided with a dial-sight, which gives the angle between the gun and any visible mark: it is practically a theodolite with the zero of its horizontal circle adjusted to the line of the gun. The method of laying the gun is then: 1. The co-ordinates of the directing gun are determined by distance and bearing from the nearest bearing-picket. From these, and the co-ordinates of the target, the battery zero-line and the range are calculated. 2. The director is set up at the bearing-picket and adjusted, by means of the known bearings from this picket, to read zero on the battery zero-line. It is then turned on the dial-sight, and the angle read. 3. The dial-sight being set to the supplement of this angle, the gun is turned until the director station comes on the sight; the gun is then on the battery zero-line. The sight is then turned on to a well-defined point—the aiming point—whose reading is recorded. The gun can then be relaid on the battery zero-line at any time, or given any desired deflection from it.

In practice there are certain complications due to the fact that the centre of motion of the gun is not the same as that of the dial-sight, but the effect of this can be eliminated by a suitable procedure, which is equally applicable to the more difficult case of the railway gun, that is trained by moving it round a curve. The only part of the operation which gives any trouble is the resection of the director position from the trig. points when no bearing-picket has been provided. Many solutions of this problem have been discussed by Captain McCaw in the *Geographical Journal* (Aug. 1918). The method given in the *Textbook of Topographical Surveying* (Close and Cox) seems on the whole the most convenient, but requires careful attention to sign.

When targets have to be found from map-detail, artillery boards are provided. These are made by cutting up maps into sections and pasting them down on a carefully drawn grid on a board of three-ply wood or zinc. Accurately drawn and checked arcs are centred upon the position of the battery and adjusted to the battery zero-line. This method of treating the map eliminates all possibility of error in scale and azimuth, and allows a battery to turn from one target to another with great precision.

Sound ranging.

If the report of a gun reaches two stations at times which differ by the quantity T, the gun must lie somewhere on the hyperbola whose foci are at the two stations, and whose semi-axis major is half the distance traversed by sound in the time T. This is the base-principle of sound ranging, and was not any novelty; but it was not developed into a working method until about the end of the first year of the war. At first the intervals of time were measured by carefully compared stop-watches; but no considerable success was achieved until the automatic registration on a photographic film was perfected. The principal features of the apparatus are: 1. Microphones, generally six at equal intervals on a circular arc carefully surveyed and referred to the standard system of co-ordinates. 2. A central station with Einthoven galvanometer wired to the microphones, so that a sound-wave reaching any microphone is registered on a travelling strip of film, with time-scale marked by a tuning-fork. 3. A forward observer, beyond the microphone base, to set the recording apparatus going as soon as he hears a report which he judges worth attention. 4. A recording officer skilled in the interpretation of the record, who can distinguish between the waves produced by different calibres of artillery, and can disentangle the true report of the gun from the disturbance produced by the scream of the shell. 5. A sound-ranging board on which the results can be worked out graphically in a few minutes, and the co-ordinates of the gun reported to the counter-battery officer. In practice it is generally sufficient to consider the hyperbolas as coincident with their asymptotes over the portions actually used. 6. An efficient meteorological service for determining the corrections due to wind and temperature. The sound-ranging sections rendered invaluable service to the artillery, and were among the great successes of the war.

It is not clear that the method has any useful application to survey on land in peace; but the analogous method of sound ranging in water may prove very valuable in hydrographic survey. As soon as the armistice was signed a channel was swept through the mine-fields from the Thames to Rotterdam, and buoyed. As each buoy was dropped a depth charge was exploded, and the position determined by the sound-ranging stations on the coast of the North Sea was reported to the Hydrographer at the Admiralty, so that a chart was ready at once, which by the older methods of hydrographic survey might have required a long time for completion in the usual November weather. It would seem that this method is well adapted to the charting of buoys or of wrecks, and the fixing of soundings.

Flash-spotting.

The principle is obvious, that the position of a gun can be determined by intersecting the place of the flash from two or more stations of known position furnished with a form of theodolite having a wide field of view and easily manipulated. The practical difficulty lies in directing the various observers to the same flash. A beautiful system was developed during the war by which all the observers of a group were connected telephonically with each other and with the central post, where a "flash and buzzer board" was installed. When the leading observer got on to a flash in a new position, he directed the other observers to it, and as they got on to the succeeding flashes from the same gun they signalled the fact to the central station. When the observer there found that they were all observing simultaneously, and therefore presumably on the same flash, he asked for the bearings from grid north, and plotted them on a mounted map showing the grid-position of each station, and deduced the co-ordinates of the gun. Success was gained by very careful training and organisation. A constant difficulty, both in flash-spotting and sound ranging, was the destruction of the electric leads by shellfire or by tanks.

Chapter X

LAND SETTLEMENT AND PROPERTY SURVEYS

In a new territory, where land belongs originally to the Government, and is sold piecemeal, the only secure method of delimiting the boundaries of property is by reference to a complete triangulation of the country. This is forcibly stated in a letter addressed by the late Sir David Gill, then Her Majesty's Astronomer at the Cape, to Earl Grey, who in 1897 was administering the Government of Rhodesia. The following extracts from this letter are taken from the introduction to the *Report on the Geodetic Survey of South Africa*, vol. III.

"The point which it is always difficult to bring home to the lay administrative mind is that it is impossible to survey a country properly or to grant indisputable titles to land by surveys made in a patchwork way.

"When Government has a particular bit of land to be disposed of, it seems to be supposed that one has only to send a surveyor to set up beacons, survey the ground, and bring back a diagram of those beacons —and sell the land according to this description.

"Assume the surveyor to be competent and honest, the result will be certain points marked on a piece of paper, representing pretty accurately the shape of the piece of land so surveyed; but in many places there will be no sufficiently well-marked topographical features, such as well-defined river boundaries, etc., to locate precisely where in the country the particular farm is; there will be nothing to indicate the latitude and longitude of any particular feature of the map (i.e. its position on a general map of the country); and there will be no accurate topography on the diagram, because the price which the surveyor receives for such survey makes it impossible for him to include accurate topography in his work. With the approximate methods of base measurement necessarily used by such a surveyor there will also be some appreciable error in the scale of the map; and beyond a rough orientation by compass (and the direction of the magnetic meridian is subject to large secular variation), there is nothing to indicate the true north of the diagram.

"Afterwards, when you come to patch such surveys together, you can see that no small trouble must arise, and that shortcomings and overlaps occur.... Then comes the further possibility, the dishonest proprietor. A particular piece of land is bought from Government. The proprietor finds a convenient spring or piece of land outside his boundary; there is no neighbour to be hurt, or perhaps only a distant proprietor. It is not a difficult matter to shift a beacon—there is no one particularly interested, and the beacon is shifted.... In this way large tracts of land have been stolen from Government in Cape Colony.

"There is one, and only one remedy for all this, and that is to connect all detached surveys with a general system of triangulation; and it will save the Government and the inhabitants generally a vast amount of money to establish this triangulation as quickly as possible."

This is counsel of perfection, and it rarely happens in practice that triangulation precedes land settlement. But it does very often happen that land open for settlement is laid out into plots by traversing with a theodolite, and this generally means that the country is cut up into squares without any regard for its topography. Thus for example the vast area of Canada under the control of the Dominion Lands department is cut up into "townships and ranges" giving squares of approximately a mile a side. The boundaries are meridians and parallels; and since in the latitudes of Canada the meridians are strongly convergent, one can get approximately equal squares only by breaking step rather often. It would seem evident to anyone born in a long-settled country such as England, where the main lines of communications (other than the modern railways) grew up in natural relation to the topography, and preceded the establishment of fixed property boundaries, that any geometrical division of the country independent of the topography must be exceedingly inconvenient, and be a permanent handicap. This is, however, not at all the view taken of the matter in Canada, where it seems to be thought that a farm is not a farm unless it is exactly a mile square, and where a railway company with land for sale found it impossible to dispose of plots laid out with what they thought would be convenient frontages to rail and river, and were compelled to re-apportion it into the conventional squares.

It remains true, however, that any such division of land in advance of accurate triangulation produces awkward misfits, and that the boundaries which are supposed to be meridians and parallels are often a good deal askew when properly mapped.

It will often happen that mining concessions and other allotments of land are made without even such control as is given by dividing the land into squares. In such cases the boundaries are traversed by a surveyor, who places a beacon at each angle, and produces a plan which is deposited in the office of the Commissioner of Lands. Such a plan is useful in documents of title, and gives the area for taxation. But an assemblage of such plans cannot make a map.

Detailed land survey.

Detailed plans of property are made entirely with chain or tape: in British territory usually with Gunter's chain of 22 yards—the length of a cricket pitch. The chain is divided into a hundred links, and ten square chains make an acre. This is interesting as one of the rare examples of a pure decimal system in British practice.

The principle of land survey is that every detail must be fixed by measuring its perpendicular distance from a straight line, and the position on that line of the foot of the perpendicular. Suppose, for simplicity, that a boundary is made up of a series of short straight lengths *AB*, *BC*, *CD*, *DE*, *EF*, ... Suppose a straight line *KL* is laid out as near as may be to the boundary, and that perpendiculars *Aa*, *Bb*, *Cc*, ... are drawn from *A*, *B*, *C*, ... to *KL*.

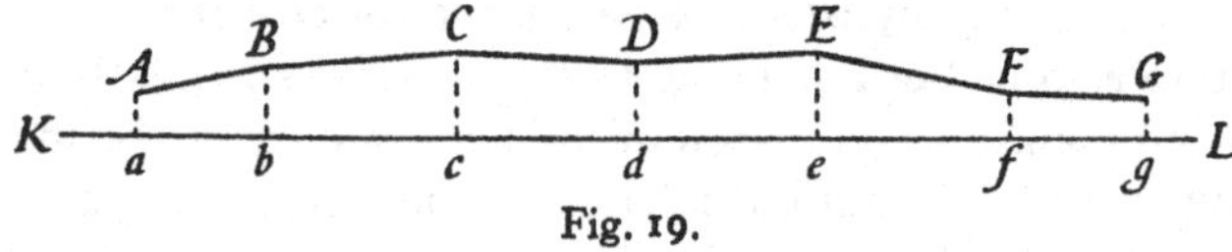

Fig. 19.

Measure the lengths of the perpendiculars, and the distances *Ka*, *Kb*, ... *KL*. A plan of the boundary *AB* ... *F* ... may now be drawn as follows: Draw the line *KL* on any desired scale, and lay off the distances *Ka*, *Kb*, ... on it. Erect perpendiculars from *a*, *b*, ... and lay off on them to scale the measured distances *aA*, *bB*, ...; we have then transferred to the plan the relative positions of *A*, *B*, *C*, ... *F*. Join them up, and we have a plan of the boundary.

The measurements along the line *KL* are made with the surveyor's chain or with a steel tape; the perpendicular distances, called offsets, are measured with a graduated offset rod, if they are short, as they should be.

When offsets are not short they must be measured with the chain or tape; and it is then necessary to have some such instrument as the cross staff or the optical square, to set out the offsets perpendicular to the chain line. But long offsets are used only in rough work, and it is a cardinal principle of accurate land survey that offsets must not be long. A chain line must pass near every point that has to be fixed. The first thing to attend to, therefore, in making a land survey, is to lay out a suitable system of chain lines.

Suppose, for example, that the plot *ABCD* is to be surveyed. The elementary, often practised, but essentially bad method, would be to chain from *A* to *C*, and take offsets on each side of *AC* to the salient points of the boundary. This would be a bad, or at least untrustworthy method, (1) because the offsets are long, and the perpendiculars must be laid out with some elaboration; and (2) because there is no check upon the result except doing it all over again.

The proper method is to break up the ground into triangles. Chain from *a* to *b*, *b* to *c*, and *c* to *a*, taking offsets to the boundary along each chain line. The offsets are now short, and are easy to measure with sufficient accuracy. Do the same from *a* to *d*, and from *d* to *c*. But now

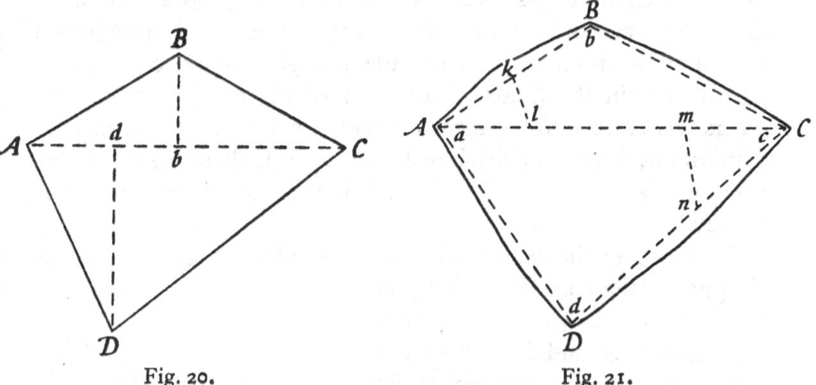

Fig. 20.

Fig. 21.

note that the construction on our plan of the triangles *abc*, *adc*, depends on the accuracy of the measures of the lengths of our chain lines. An error in one would completely wreck the whole plan. Therefore they are to be checked by measuring the lines across the corners *kl*, *mn*. When the plan is drawn out, these distances, as measured on the plan, must agree with the actual distances measured in the field. If they fulfil this condition, then we have a complete check on the main structure of the work. If they do not, the fact that there is an error somewhere is detected at once.

It has been mentioned already that a plan does not generally show the inequalities of the ground. It is one of the principal objects of a map to do so. But whether the inequalities are indicated or not, the plan or map must indicate the relative positions and distances of all objects as if they are projected by vertical lines on to a level surface. Distances must be measured and represented as horizontal distances, and the distinction is by no means insignificant in the case, for instance, of property on the side of a hill. In all land surveying by chain and offset rod, therefore, the chain and rod must always be held horizontal; and the horizontal equivalent of a distance down a slope will be measured in a series of steps. When a slope is steep this becomes difficult, and it is evident that the chain lines must be chosen so that they are as level as may be.

In theory land surveying is exceedingly simple, consisting only in measuring along the sides of the triangles, and taking offsets to all detail. The whole art of it in practice lies in selecting the system of chain lines so that they shall form as rigid a framework of triangles as can be made. If the three sides of the triangle are known with absolute precision the triangle can be constructed rigidly, within the limits of accuracy of the draughtsman. But errors will creep into the measurement of sides, and the draughtsman cannot draw with absolute accuracy. It is therefore necessary that the triangle shall be "well conditioned", as nearly equilateral as may be.

For consider the construction of a triangle with an acute angle *C* opposite the base *AB*. The point *C* is to be found by describing circles with centres *A* and *B*, and of radii equal to the measured lengths *AB*, *AC*. These circles will intersect at an angle equal to *C*, and it is evident that the more acute the angle, the greater will be the error in the position of *C* caused by an error in one of the measured lengths.

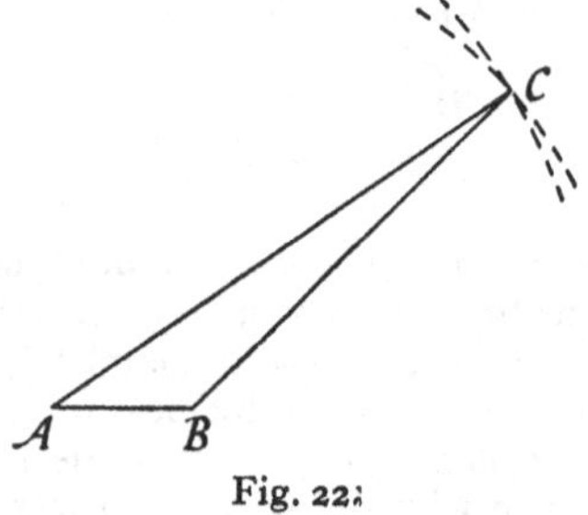

Fig. 22.

The same kind of argument applies in all branches of surveying that depend upon triangulation of any kind: that is to say, in all kinds of survey except the simple route traverse. There must be no unduly acute angles in the triangle.

The method of chain survey is well adapted to making plans of small detail. The plan of each triangle, and of the points fixed to it by offsets, has considerable accuracy in itself. But it will be understood easily that a large estate cannot be surveyed in this way, in a number of small triangles, because every triangle is inaccurate to a small degree, and the effects of these small errors rapidly accumulate when an attempt is made to fit a large number of the triangles together. In this, as in all branches of survey, it is impossible to make an accurate extensive map or plan by piecing together small portions surveyed independently, to build up a large map from small blocks. To take the exactly opposite way is the right course. Construct a simple framework covering the whole extent of land to be surveyed; get this framework so accurately made that it cannot be in error by an amount visible to the eye on the scale which is proposed for the map; and hang the detailed survey upon this framework. An accumulation of error is then impossible.

The large-scale maps of the Ordnance Survey.

In accord with this principle the large-scale surveys of a National Survey such as the Ordnance Survey are based upon the general framework of triangulation, which is broken down into secondary and tertiary triangulation, the sides of a tertiary triangle being not much more than a mile. All the detail is then tied to the triangulation by chaining. We have already remarked (page 11) that these maps, though commonly called cadastral, are not properly so called, first because they show the visible impediments to trespass and not the true boundaries of the property, and secondly because there is not in England any system of recording the coordinates of boundary marks in terms of the origin of triangulation. The description is graphical, not numerical.

Chapter XI

GEODETIC SURVEY

THE present meaning of the word Geodesy is the same as the old meaning of the word Geometry: the measurement of the Earth. And it is a great pity that the word is sometimes used, in examination schedules, to mean elementary survey.

Figure of the Earth.

The principle which underlies the measurement of the Figure of the Earth is exceedingly simple.

Suppose first that the Earth is, to the best of our knowledge, spherical; and consider how we should measure the radius of the sphere. Take two stations on the same meridian, and determine the latitude of each. This gives the distance between the two stations in angular measure on the sphere. Now, by triangulation, measure the distance between the two stations in terms of the unit of length that we prefer, say metres. We then have the result that so many metres are equivalent to so many degrees, minutes, and seconds of the arc on the sphere. If the difference of latitude is n'', and the distance between the stations along the meridian is M metres, then the radius of the meridian in metres is given by the equation

$$\text{Radius} = M \operatorname{cosec} 1''/n.$$

This is the principle of the method used by Eratosthenes in his celebrated attempt to measure the Earth, in the third century B.C.

If similar operations carried out in many different latitudes on various meridians gave always the same value for the radius, we should have the clearest evidence that the Earth was a true sphere. But in fact we get different values for the radius when we determine it in different places. This was suspected in the latter part of the seventeenth century; but the errors incidental to early operations of the kind led to some confusion. The observation of Richer at Cayenne in 1672, that the pendulum which he had adjusted in Paris required shortening to keep time near the Equator, supported the theory enunciated by Newton and established by Clairaut, that the rotation of the Earth must involve a flattening at the poles. On the contrary, the observations on an arc of meridian in France led Jacques Cassini to conclude the opposite. The celebrated expedition sent by the Academy of Sciences to Lapland enabled Maupertuis to "flatten at once the Poles and the Cassinis" and his results were confirmed by the other expedition sent by the Academy to Peru. By the middle of the eighteenth century the law was definitely established for the northern hemisphere: the further one goes north, the larger is the distance corresponding to a degree of latitude; the flatter, therefore, is the Earth.

The measures made in Peru and in Lapland, combined with those made in France, were consistent with the hypothesis that the Earth is a spheroid of revolution, or, in other words, that all the meridians are similar ellipses. They were consistent with this idea, but they were by no means sufficient to prove it; and in the middle of the eighteenth century grave doubt was cast on the matter by the result of the Abbé de Lacaille's measure of an arc of meridian at the Cape of Good Hope, which seemed to show that the southern hemisphere was prolate, the degree decreasing in length as the pole was approached.

It was some time before the explanation of this difficulty was discovered; and in the meanwhile a somewhat similar discrepancy had been found in England, the curvature of the southern half of England appearing to be less than that of the northern half.

Deviations of the vertical.

The origin of these abnormal results is in local deviations of the vertical, which have been mentioned already in Chapter IX, page 180. The latitude of any station is the inclination of the Earth's polar axis to the horizontal plane of the station. The horizontal plane is the plane at right angles to the direction of gravity. The direction of gravity is determined by the distribution of matter within the crust of the Earth. If this distribution is abnormal in the neighbourhood of a station the horizontal plane is no longer a tangent plane to the spheroid which represents the general form of the Earth; and if such a place is chosen by bad luck as one of the terminals of an arc of meridian the curvature of the meridian deduced from these measures is not representative of the general average curvature of a meridian in that latitude.

We have said that the direction of the vertical is influenced by the distribution of matter within the Earth's crust. It is also of course affected by the attraction of the visible mountain masses above the level surface. But these latter can be allowed for; and when this is done it nearly always happens that the visible masses prove to be quite insufficient to account for the results obtained

At Moscow, far from any considerable mountain masses, there are rapid changes in the deviation of the vertical which cannot be accounted for by anything above ground. In India, on the other hand, the very visible masses did not produce anything like the effect that they should.

The attraction of the Himalaya.

The problem of the effect of the Himalaya on the direction of gravity in India has been present continually to the Indian surveyors. North of the Indian arc is the immense mass of the greatest mountain range in the world and the high plateau of Tibet; to the south the Indian ocean is very deep, and the consequent deficiency of density considerable. It might be expected that in India the direction of gravity would be deflected towards the north; and calculation shows that the effect of the visible excess of density to the north, the visible deficiency to the south, should extend all over India.

A great number of careful determinations of latitude were made at stations along the great arc of meridian, but the expected effect was not obtained. The discordances between the astronomical and the geodetic latitudes were considerable; but they did not fit in at all well with the deviations which the visible masses should have produced. It appeared that some cause was at work which in great part annulled the effect of the mountain attraction. This was the first indication of the law which now plays so large a part in these enquiries, that in some way the expected effect of mountain masses is compensated, as if there were underlying deficiencies of density which nearly balanced the visible masses.

About the year 1860 Archdeacon Pratt, of Calcutta, a distinguished mathematician from Cambridge, applied himself to the mathematical investigation of this problem, and arrived at the above result, that the attraction of mountains is not so great as their visible masses would lead one to expect. The then Astronomer

Royal, Sir George Airy, proposed to explain this in the following way: Conceive that the Earth is composed of a solid crust about forty miles thick, with a liquid below. The strength of the crust would not be nearly sufficient to support the weight of the superincumbent mountain masses; therefore there must be some support for them, and this may be in the form of protuberances beneath the mountains, of the lighter crust into the denser liquid below. In such a way the mountains would be in equilibrium because they are practically floating like icebergs in the ocean, buoyed up by the intrusion of their "roots" into the liquid.

This "roots of the mountains" theory has been the subject of much interesting discussion, especially by the Reverend Osmond Fisher, in his *Physics of the Earth's Crust*. Mathematicians have in general felt themselves compelled to reject the internal fluidity of the Earth, on account of certain tidal phenomena, into which we can hardly enter. But the important idea that the principal mountain masses are nearly in equilibrium, their visible excess of weight balanced by invisible defect of density below, as if they were floating, has by gradual steps become an established principle of geodesy.

If the visible excesses of dense material in the mountains are counterbalanced by deficiencies underneath, it is clear that we may expect also that the visible deficiencies of matter in the oceans should be balanced by excesses of matter in the ocean beds. Evidence for this supposition cannot be found however in the observations for latitude, so well as in the conclusions to be drawn from the determinations of gravity by the pendulum, to which we must now refer.

Gravity survey with the pendulum.

The mathematical theory of the attraction of a spheroid at any point of its surface, or at an external point, provides a formula which expresses the force of gravity, and thence the time of oscillation of a pendulum of known length, at any place. Knowing the shape of the Earth, we can predict the rate of swing of the pendulum; and alternatively, it is evident that if we carry an invariable pendulum about the world, and determine the time of its swing at different places, we have a means of determining the ellipticity of the Earth independent of the triangulation and latitude method explained above. When sufficient observations were accumulated it was found that the two independent methods of arriving at the ellipticity of the Earth gave results which were in tolerable agreement, though they were not identical. For the moment the discrepancy need not concern us.

But it was also found that the pendulum observations on mountains gave peculiar results. It is easy to calculate the intensity of gravity at a given height above the surface of the Earth, and to allow for the increase which should be due to the mass of the mountain on which the pendulum is established. But observation discloses the remarkable fact that the mountain mass itself has not the expected effect; the pendulum swings more nearly as if it were somehow supported at the height of the mountain, but unaffected by the mountain mass. In other words, the attraction of the mountain is more or less compensated.

Further, it was pointed out by the French geodesist Faye, that a similar effect is found on oceanic islands: similar in cause, though opposite in its immediate effect on the time of swing of the pendulum. On those volcanic islands which rise steeply from deep ocean the force of gravity, as determined by the pendulum, is generally in excess, the excess being about that amount which can be attributed to the mass of the island standing above the ocean floor. In other words, if the pendulum could be swung on the surface of the sea, without the material support of the island, then the time of swing would be normal.

The determination of gravity at sea.

The first attempt to verify this idea was by Hecker, who made several long voyages, including a circumnavigation of the world, comparing all the while the readings of a large number of mercurial barometers and boiling-point thermometers. Both these instruments determine the pressure of the atmosphere, but whereas the former is affected by the intensity of gravity the latter is not. Similar work was carried out later by Duffield, with more instrumental refinements; and both obtained the general result that gravity over the deep oceans is more or less normal.

A great advance was made when Dr Vening Meinesz of the Dutch Geodetic Service constructed a geodetic pendulum apparatus which could be used at sea in a submarine. During a voyage

Plate XVII

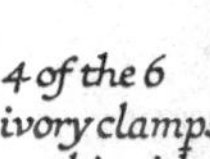

(*Geogr. Jour.* LXXI, 157, *February* 1928)

Ieinesz Geodetic Pendulums for Determination of Gravity at sea in a submarine.

1. *Barr and Stroud Topographical Stereoscope.*

(*From paper by Capt. M. Hotine, R.E. Geogr. Jour.* LXXV, 144, *February* 193

2. *Contoured strip and finished plotting: Arundel method.*

from Holland by Suez to Java in 1923–4 he obtained results which gave very small residuals from the normal, and showed that the Indian Ocean is very nearly perfectly compensated. But Dr Meinesz then made a second voyage to the Dutch Indies, via the Panama Canal, of which the results show excess of gravity over very large areas both in the Atlantic and the Pacific. It seems likely, therefore, that as usually happens, the matter is not so simple as it appeared at first sight.

The form of the ocean surface.

In connection with this subject of mountain attraction it is natural to speculate what effect is produced upon the form of the ocean surface by the attraction of great mountain masses. Several calculations have been made which show that if the mountains can exert the full effect that their visible masses entitle them to, the elevation of the free surface of the ocean must be considerable. Colonel Clarke (*Geodesy*, page 94 et seq.) gives a method of calculating the order of the effect upon the height of the sea surface produced by the attraction of the Himalayas, and shows that it might amount to as much as 600 feet. "This calculation," he proceeds, "shows us that large tracts of country may produce great disturbances of the sea level, but it is at least questionable whether in point of fact they do." The compensation of the mountain masses nullifies the effect in great part, we may be sure, and leaves little scope for the existence of great differences in the sea level radii of the Earth.

Nevertheless it is common to meet in books the statement that the sea level at Calcutta is about 330 feet higher than it is at Cape Comorin; and that the sea on the Pacific coast of South America, close under the Andes, is 2000 feet higher than it is at the Sandwich Islands. These statements are based on calculations such as Colonel Clarke gives; but they are sometimes said to be supported by the results of direct levelling. Such a statement is on the face of it absurd. The surface of the sea must be a level surface in the sense that spirit levelling from one point to another could give no evidence of rise or fall. Any attraction of mountain masses which changed the form of the sea surface would have an equivalent effect on the results of the instrumental levelling.

The surface of the sea does however at least in one region depart considerably from the spheroid. Dr Meinesz has recently found in the Netherlands East Indies a line of large "negative anomalies" which indicate a depression in the sea surface of some 30 or 40 metres, making, curiously enough, a kind of boundary ditch between Asia and Australasia.

The theory of "isostasy".

The convergence of various lines of investigation towards the same point seemed to establish a general law of balance: where there is excess of mass above it is more or less balanced by deficiency below, and vice versa. To this condition the American geodesists have given the name Isostasy, proposed in an important paper by Major Dutton, though it is far from clear that what he meant by the word is the same as what is now implied.

This celebrated theory was greatly developed by another American geodesist, the late Mr J. F. Hayford, of the United States Coast and Geodetic Survey, in two very elaborate investigations published in 1909 and 1910. He showed that the calculation of the attraction of visible masses must be extended over the greater part of the Earth and devised an elegant way of doing this without undue labour; and he followed Pratt in the artificial assumption that the compensation becomes complete at some definite depth, irrespective of the mass to be compensated. On this hypothesis the mass in a column of unit area extending from the visible surface, either land or sea, down to the "depth of compensation" would be everywhere the same; and for convenience of calculation he assumed, moreover, that the compensating excesses or defects of density were distributed uniformly through the column. These artificial assumptions have proved something of a stumbling-block, since it is difficult to imagine any cause which would produce the necessary adjustments of material to establish and maintain such a condition. More recent investigators have somewhat modified the assumptions, while retaining the general idea that there is such a thing as one depth at which compensation is complete. The theory of the figure of the Earth is thus much entangled with assumptions that are defensible only on the ground that they provide working hypotheses that can be investigated numerically. To any such investigation a formal objection is sometimes raised, that sounds formidable. There is a well-known dynamical theorem due to Stokes, that it is impossible anywhere outside an equipotential surface to discover the distribution of mass within. Sea level is an equipotential surface: therefore no observations of deviations of the vertical or abnormalities in the value of gravity made above sea level can determine the constitution of the earth below. This is formally true. If it were also practically true, there would be an end of all investigation of what produces this compensation of the visible mountain masses. But there is also a good principle that of a multitude of possible causes one is entitled to choose the simplest. If then one can by trial determine a simple arrangement of masses below the surface which accounts for the facts observed above it, one may rightly prefer that solution to others of the infinite number formally possible but physically unlikely.

The Pratt-Hayford theory of compensation is losing ground, and Dr Jeffreys' investigations of compensation are all now based upon the flotation theory of Airy.

With this descriptive introduction to the present state of the theory of the figure of the Earth, we are in a position to discuss briefly the methods and results of modern geodesy.

Geodetic triangulation.

The first test of accuracy of triangulation is the size of the triangular error: that is, of the average deviation of the sums of the measured angles in the triangle from 180° after they have been corrected for spherical excess (see page 191). In geodetic triangulation the triangular error must not exceed one second, and in the best modern work does not exceed 0".75. To attain this great accuracy the following precautions must be taken:

The theodolite must be of the finest construction, though not now necessarily of large size (see page 265). Its precision must be verified with the greatest care, and every precaution taken in its use in the field. It must be mounted on concrete pillars at each station, and scrupulously adjusted over the station mark.

The signals should generally be lamps, electric or acetylene, and all the observations made at night, when the refraction is steadier.

The positions of the stations must be carefully reconnoitred, to avoid grazing rays that are subject to unknown horizontal refraction. The best practice now avoids rays of extreme length, on which observation may be interrupted by bad weather for long periods, and which often require that the signal shall be carried on a high scaffold much above the theodolite station, leading to difficulties of centering. It is more economical to employ smaller triangles and more of them.

The rounds of angles must be repeated on different zeros until the means are accurate enough to give the required average of triangular error. To be satisfied that the results are coming out of this accuracy, it is necessary to carry on a provisional reduction of the observations in the field, so that anything abnormal may be detected at once, and remedied before the station is left far behind. This requires a strong party.

Modern practice simplifies the calculations by rejecting the old idea that a measured angle should be weighted according to the degree of concordance of the individual measures of it. Systematic errors such as those caused by horizontal refraction are more dangerous than accidental, and the concordance of measures is no guarantee against it.

The geodetic triangulation of Great Britain was a network covering the whole country. More modern principal triangulations have usually taken the form of chains of quadrilaterals running along meridians and parallels to form a grid. We shall find reason to think that the strong distinction hitherto drawn between geodetic and other triangulation is likely to disappear, and that in the future all careful triangulation will be made to contribute to the problem of the figure of the Earth.

When a geodetic triangulation is adjusted to close on a series of bases at distances of one or two hundred miles, its length is very rigidly controlled; but it is always possible that the small remaining errors in the observed angles will give the chain a twist. Astronomical observations of azimuth would serve as a good control of such twist were it not for the fact that azimuths, like latitudes and longitudes, are affected by local deviations of the vertical. There is however a relation, due to Laplace, between the effect on longitude and on azimuth, so that if the longitude term can be found; the correction to the observed azimuth can be deduced. Thus at any "Laplace Point"—that is, at any point where astronomical longitude and azimuth have been obtained—the azimuth can be corrected and terms can be introduced into the equations of condition for the adjustment which closes the triangulation upon all the astronomical azimuths as well as on all the bases. This very important improvement in modern geodetic practice is of course due to the ease and accuracy with which longitudes may now be obtained from wireless time signals.

Geodetic bases.

In the chapter on trigonometrical survey we have dealt with the use of invar wires in the field, for the measurement of bases. We will mention now some of the refinements that are necessary when the work is to be of the highest precision.

At the headquarters of an important geodetic survey provision will be made for the standardisation of the wires, to ensure a more complete control than is possible when they are merely returned to some far distant laboratory at intervals.

First it is necessary to provide a standard bar, which should be the unit of length, yard or metre, multiplied by some power of two, so that it can be compared by successive duplication. Four metres or four yards will be the most convenient length. But in fact, ten feet has been the usual length of English bars, though this involves an awkward comparison with the British standard yard. It is not quite clear why, when the yard is the standard unit of length, British geodetic measurements have been made in feet. The modern bar will be made of invar. It will be compared with the standard metre at the International Bureau of Weights and Measures at Breteuil, or with the standard yard at the Board of Trade Standards Office, or with the copies of this standard at the National Physical Laboratory or at the Ordnance Survey Office, Southampton. At the same time the laws of its expansion will be minutely studied.

To avoid duplication of statement we shall speak of the operations in future as conducted in metres.

The next step is to establish a 24-metre comparator at headquarters, standardised from time to time with the 4-metre bar. This comparator will be in the form of a wall, preferably underground to escape temperature changes as much as possible. It will be provided with tanks in which water can be circulated, to allow for the study of the temperature coefficients of the wires.

On this comparator the wires will be standardised before they go into the field, and when they return.

The constancy of the wires depends very much upon the care with which they are treated in the field. They must always be wound upon the special aluminium drums which are made for them, and they must be treated as instruments of precision should be treated, not after the fashion of coils of wire. One of the principal difficulties in the field has been to ensure continuous careful treatment for the wires, and it has sometimes been suggested that if they cost more they would be treated with more respect. This remark was made in pre-war days, when an invar tape cost thirty francs; but they are still a relatively inexpensive part of the geodetic surveyor's equipment.

When they are first manufactured they are subject to molecular changes and their lengths are not constant. As they become aged they settle down into stability, and the process may be quickened to some extent by successive very careful annealing. But it does not seem that anything can complete the natural process of ageing except use in the field.

It is essential that the coefficient of expansion should be found for each separate wire. At first it was the practice to test a sample of the rolling from one ingot, and to assume that all the batch had the same constants. It is now recognised that this is not safe and that each wire must be examined after it has been made up into its working form.

There is still some difference of opinion as to the relative merits of wires and tapes. The advantage of wire is that it is less subject to the disturbing influence of wind. Tapes have several advantages; twist can be detected very easily; they are not so liable to kink; and the small divided scale can be engraved on the tape itself, instead of on a soldered attachment, the "reglette". The last is of considerable importance.

Experience of their use in rough country has shown that the wires or tapes can be used on much greater slopes than was considered desirable when they were first introduced. But this requires that the provision for determining the slope of the tape shall be more thorough than was made in the first patterns. Experience on the Semliki base in Uganda, and on the Lossiemouth base in Elgin, favoured the use of an ordinary Y level, and a special light levelling staff which can be stood on the tripods carrying the fiducial marks against which the tape is read. With this equipment it is possible to measure up slopes of 1 in 3, and to choose for the ends of the base situations which provide a good view of surrounding stations favourable for the base extension.

This involves a thorough discussion of the effect of slope on the horizontal distance between the end marks of the tape, arising from the change in the form of the catenary, and from the difference of tension at the upper and lower ends—a difference which may become so considerable that the pulleys must no longer be frictionless, or the tape will run away down hill. A very complete investigation of the problem has been made by Professor Henrici and his son Captain Henrici, R.E.; the results are too complex for summarisation here. See Ordnance Survey: *Professional Papers*, New Series, No. 1.

For rapid work it is essential that the base party shall be well drilled. With a well-trained party a base of 10 km. can be measured completely in 12 days; and it is good economy to arrange that all the bases required for the whole triangulation shall be measured consecutively in a single season if possible.

The astronomical observations.

The observations for latitude have usually been made by Talcott's method. Pairs of stars are chosen which transit at nearly equal meridian distances north and south of the zenith of the station. With a zenith telescope, or a theodolite adapted to serve as such by fitting it with an eyepiece micrometer, the difference of zenith distance at transit is measured by the micrometer, and thence from the known apparent declinations of the stars the latitude is deduced. Since the zero of the zenith telescope is determined by level, the concluded latitude is affected by the local deviation of gravity.

The observations for time and longitude will be made with portable transit instruments, fitted with the travelling wire micrometer to eliminate as far as possible personality in the observation varying with the magnitude and apparent speed of the star. The transits will usually be recorded on a field chronograph, which may also by suitable relays be made to record the wireless time signals: though it is not certain that this complication of field apparatus is really necessary or desirable, and excellent results can be obtained by various devices for comparing the chronometer and the rhythmic time signals by listening with a telephone headpiece.

Those who have used the prismatic astrolabe in the field are unanimous in preferring it to the theodolite; and it seems likely that with larger and improved astrolabes the zenith telescope and portable transit instrument will also be superseded.

The azimuths must however still be observed with the theodolite, generally by the method of circumpolar stars at maximum elongation.

We shall see in what follows that far more astronomical observations are required in geodesy than have been made hitherto, and that many stations observed with moderate accuracy are more valuable than fewer with a degree of refinement that proves to be superfluous.

Determination of the figure of the Earth.

We have now a rigid framework, whose dimensions are rigidly controlled in terms of our standard of length; which is pinned down, so to speak, at an initial latitude, and in azimuth at the Laplace points. By the aid of tables calculated from some adopted figure of the Earth we have calculated the geodetic latitudes and longitudes of all the stations of the triangulation, and are now in a position to see how well our observations agree with the adopted figure.

When the astronomical latitudes and longitudes are compared with the geodetic, they will be found to differ by amounts which average several seconds of arc. These discordances are due primarily to the deviations of the vertical, due to irregular distribution of density within the Earth or to the attraction of the visible masses about the surface; and secondly they are due to the imperfection of the figure of the Earth upon which the geodetic tables were calculated. The more or less haphazard distribution of the first will tend to mask the regular run of the second. But if we can make enough astronomical observations at many stations to smooth out the irregular deviations of the vertical, we have in the systematic residuals material for improving the assumed figure of the Earth and the geodetic tables. It is not hard to form an expression which shows what difference a given change in the size and the flattening of the Earth will make in the geodetic positions. Each comparison Astronomical minus Geodetic then gives an equation relating these differences to the assumed figure; and a solution of all the equations shows what change must be made in the assumed figure to fit the observations as well as may be.

The geodetic measures which contributed to the historic determinations of the figure of the Earth are singularly few. They are:

1. The Arc de Pérou measured by the French Academicians Bouguer and La Condamine in the middle of the eighteenth century.

2. The French arcs measured by Delambre and Mechain about the end of the eighteenth century, and by Biot and Arago early in the nineteenth.

3. The British arc from Dunnose in the Isle of Wight to Saxavord in the Shetlands. When 2 and 3 were joined by triangulation across the Straits of Dover they made the Anglo-French arc of meridian.

4. The great Arc of Meridian of India.

5. The Russo-Scandinavian arc measured under the direction of the elder Struve from Hammerfest to the mouth of the Danube.

6. Various triangulations in Central Europe by Schumacher, Gauss, and Bessel.

7. Maclear's arc at the Cape of Good Hope.

Everest's figure was determined in 1830 essentially from 1, 2, 3, and 4.

Bessel's figure was determined in 1837 (corrected 1841) from 1, 2, 3, 4, 5, and 6.

Airy's figure was determined in 1849 from 2, 3, 4, 5, 6, and from Mason and Dixon's arc in Pennsylvania.

Clarke's figures of 1858, 1866, and 1880 were determined from slightly different combinations of 1 to 7, with Indian longitudinal arcs.

The only modern figure is that published by Hayford in 1909 and revised with additional material in 1910. It depends entirely on the geodetic measures of the United States Coast and Geodetic Survey corrected for isostatic compensation.

Everest's measures were expressed in terms of the Indian 10-foot bar; Bessel's in metres, but depending on a comparison of the Prussian toise with the old Toise de Pérou, and thence with the Metre of the Archives; Struve's measures were in terms of a standard toise compared with the Prussian. Clarke depended upon a comparison between the Ordnance Survey 10-foot bar with the metre via the Russian toise. There are therefore some delicate, difficult, and partly insoluble questions arising when one tries to compare one result with another. But the following figures, taken from the Introduction to my new Geodetic Tables, in which there is some discussion of the problem, will serve our purpose, though it must be confessed that their relation with the International Prototype Metre is somewhat uncertain.

	Equatorial semi-diameter a (metres)	Flattening. $(a-b)/a$
Everest 1830	6 377 304	1/300·8
Bessel 1841	6 377 397	299·2
Airy 1849	6 377 542	299·3
Clarke 1858	6 378 294	294·3
Clarke 1866	6 378 206	295·0
Clarke 1880	6 378 249	293·5
Hayford 1910	6 378 388	297·0

All these determinations were made on the assumption that the figure of the Earth is a spheroid of revolution. In 1878 Clarke had discussed the material on the assumption of three unequal axes, but it proved to be insufficient.

The whole calculations of the Survey of India are in terms of Everest's figure. The Ordnance Survey uses Airy. Most countries of Europe use Bessel, but France some time ago adopted Clarke 1880. The United States uses Clarke 1866. The tables in the official *Textbook of Topographical Surveying* are on Clarke 1858. Gill's 30th meridian Arc in South Africa was computed on Clarke 1880.

There is thus plenty of variety in the use of geodesists, which makes much trouble when their results are to be compared. For their own internal purposes one figure is about as good as another, and no one could contemplate revising the whole calculations of a survey department to bring their adopted figure of the Earth up to date. But it had long been felt that there would be some considerable advantage in agreeing that the principal results, and especially the comparisons of geodetic places with astronomical, should be referred to some one figure adopted as standard.

The above table shows that there is much more diversity in the determination of the flattening than of the equatorial radius, and we shall see later that this is easily explained, since long arcs in high latitudes are required to distinguish clearly and directly between corrections to the radius and the flattening. Helmert, indeed, had expressed the opinion so long ago as 1901 that the flattening of the Earth could be better determined from variations of gravity observed with the geodetic pendulums than from arcs of meridian.

Gravity survey.

The acceleration due to gravity varies with the latitude, since the observer on the equator of a spheroid is further from the centre of attraction than one in higher latitudes, and also because the centrifugal acceleration due to the rotation of the Earth is greater at the equator. By theory due to Clairaut the relation between gravity g at sea level in latitude ϕ and gravity g_0 at sea level on the equator is

$$g = g_0 (1 + a \sin^2 \phi + b \sin^2 2\phi)$$

b is a small coefficient determined from theory and independent of the flattening, while a is related to the flattening by the equation

$$a = \cdot 0086432 f$$

From a discussion of the gravity observations then available Helmert found in 1901 that the flattening came out 1/298·3. And in a later discussion published in 1915 he concluded that the pendulum observations gave strong evidence that the Earth is ellipsoidal, with its longest equatorial radius in longitude somewhat west of Greenwich.

A few years ago, then, the position was as follows:

There had been no general discussion of the figure of the Earth since Clarke had published his *Geodesy* in 1880—Hayford's figure was derived from measures made only in the United States. Since 1880 very much geodetic work had been done in Europe, in South Africa by Gill, in Spitsbergen by a joint Russo-Swedish enterprise, in Ecuador by the Service Géographique in remeasuring and extending the Arc de Pérou, and in many other places. There were also the gravity surveys to be taken into account. The establishment of the theory of isostasy demanded that much of this work should be recomputed with the elaborate corrections required by that theory. At the first meeting of the new International Union of Geodesy and Geophysics, held at Rome in 1922, it was resolved that a standard figure should be adopted for use in the publication of geodetic results, and that a decision should be taken at Madrid in 1924. At an Afternoon Meeting of the Royal Geographical Society on 12 May 1924, Captain G. T. McCaw, of the Geographical Section, General Staff, from a careful general examination of all the material, concluded that the best approximations to the two principal constants of the Earth's figure are:

$$a = 6\,378\,300 \text{ metres}; \qquad f = 1/296.$$

In the course of the discussion following this paper it occurred to the author of this book to propose the calculation of a new kind of geodetic table. Existing tables gave the radii of curvature in different latitudes. None of them gave completely the lengths of the arcs of meridian from the equator to different latitudes; and there was no ready method of finding how these arcs varied with the radius and the flattening.

New geodetic tables. (R.G.S. Technical Series, No. 4.)

The new tables calculated soon after this discussion took place are based on Clarke's figure of 1880. The first table gives the length of the meridian arc from the equator to each ten minutes of latitude, together with the effect upon the arc of a change of one per cent. in the flattening. Fig. 24 based upon these tables shows how the lengths of the arcs from the equator vary with the flattening and with a change in the radius of the equator. Call the first set "flattening curves" and the second set "radius curves". The former are very nearly straight between latitudes 0° and 30°; they rise to a maximum about latitude 54°, and the chords from 40° to 70° or from 45° to 65° are horizontal. The latter are on the scale of any plotted figure indistinguishable from straight lines.

Now any curve drawn among a series of points representing meridian deviations from any assumed figure should approximate to a curve compounded of a flattening curve and a radius curve, and if we can disentangle the two components we can see at once how the figure of the Earth which best represents our observations differs from that assumed in the calculations. It is clear

1. That in low latitudes the compound curve must be so nearly straight that the components cannot be distinguished. Hence a meridian arc measured in low latitudes cannot separate corrections to the flattening and the radius.

2. That an arc in middle latitudes between 30° and 50° may begin to show a curvature due to the need for a change in the adopted flattening, but that no good separation from radius correction is possible.

3. That the most useful single arcs are in high temperate latitudes between 40° and 70°. The slope of the chord gives the correction to the

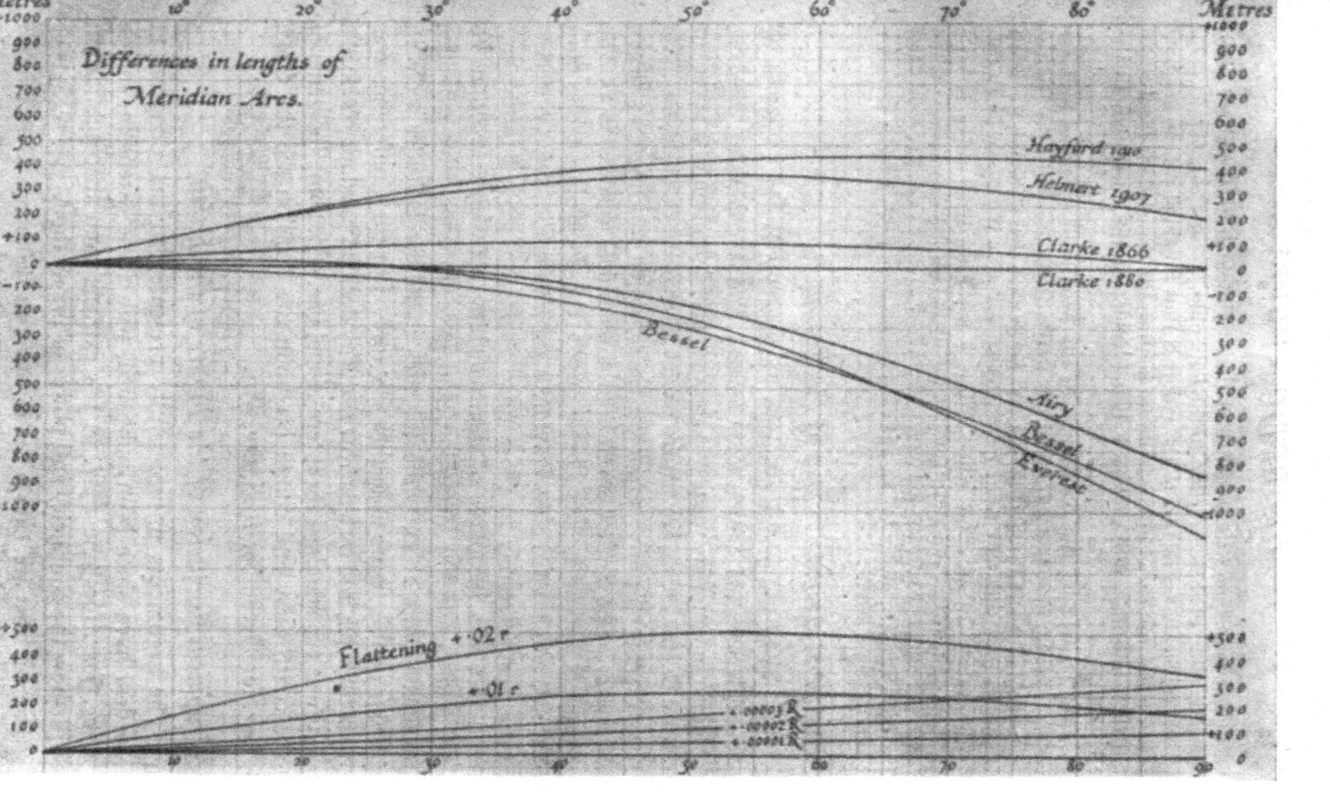

Fig. 23. *Differences of lengths of Meridian Arcs calculated on the principal Figures of the Earth.*
Fig. 24. *Effects of changes in Radius and Flattening.*

assumed radius, and the distance of the maximum from the chord gives an indication of the correction to the flattening.

4. When the correction to the radius is thus obtained we can see how far the arcs measured in low latitudes confirm the conclusions from those in high latitudes, and deduce a better value for the correction to the flattening.

Now let us plot the differences in the lengths of meridian arcs for several of the principal figures of the Earth, compared with the lengths tabulated on Clarke 1880 (Fig. 23). We see at once that although Everest's figure is very different from Clarke 1880, yet over the range of Indian latitudes—8° to 30°—for which Everest has been used there is little difference between the meridian arcs calculated on the two figures; while over the range of latitude for which Clarke's figure of 1866 has been used in the United States—25° to 50°—the differences from Clarke 1880 are very nearly constant.

The International Figure of the Earth.

The most important question for decision at the Congress of the International Union of Geodesy and Geophysics, held at Madrid in October, 1924, was the proposed adoption of a standard figure of the Earth, for future use in all general discussions. The Executive Committee proposed the adoption of Hayford's figure of 1910. The author submitted a graphical discussion of the principal meridian arcs, based upon the foregoing tables and theory, and argued that Hayford's determination, from observations made entirely in the United States, between latitudes 25° and 50°, was incapable of distinguishing between corrections to the radius and the flattening, and gave no conclusive evidence against Clarke 1880, while the arcs in higher latitudes were in favour of this figure as against Hayford. But the Conference was unconvinced, and by a narrow majority decided to adopt Hayford's figure as the standard for international use. For some incidents of the controversy and comments upon the decision see the *Geog. Jour.* for December 1924, LIV, 479.

Graphical discussion of the Figure of the Earth.

The arguments unsuccessfully presented to the Congress at Madrid have since been published, with some extension, in the *Geog. Jour.* for June 1927. A few extracts from that publication will show the simplicity and power of the method.

Suppose that we have a set of meridian deflections G–A referred to Bessel's figure. Let the deflections in seconds of arc, as ordinarily given, be turned into metres, and plotted against the curve for Bessel in Fig. 25. We can see at once, by inspection, how they will fit any other of the figures whose curves are plotted. We draw a median line or curve as best we may through the plotted points and judge of the fit to other figures by the approximation to parallelism with their curves. For example, in Fig. 25 we have plotted the meridian deflections deduced by Schumann for the Franco-British arc against the curve for Bessel's figure, on which they were reduced. We see at once that they do not fit Bessel: but that a median line drawn through them is practically horizontal: that is, they do approximate much more closely to Clarke 1880.

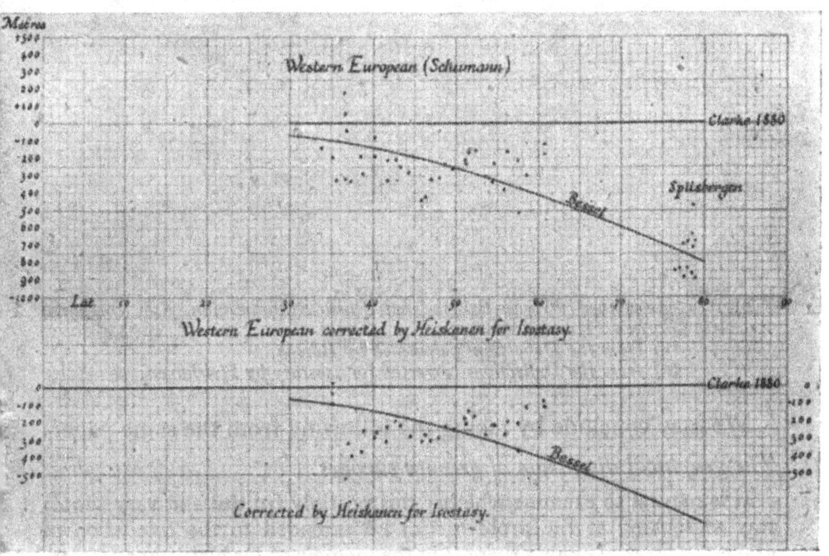

Fig. 25. *The Western European Arc compared with Bessel's Figure (a) from the Deflections calculated by Schumann, (b) from the Deflections corrected for Isostasy by Heiskanen.*

The combination of the Franco-British and Russo-Scandinavian arcs by Börsch (Fig. 26) gives much the same. If the median curve is not a straight line its convexity is downwards. We conclude that these arcs indicate no increase in a, and if anything a decrease in r.

From these and other comparisons (for which we must refer to the original paper) one may conclude that the principal geodetic arcs fit Clarke 1880 as well as they fit any other figure, and that this result is not sensibly changed when isostasy is taken into account. And the graphical method also shows very clearly that the material is at present insufficient to give any conclusive result. The random deviations of the vertical are so large that the relatively few astronomical latitudes are quite insufficient to discriminate certainly between one figure and another, and a principal need of geodesy is to observe at least ten times as many latitudes.

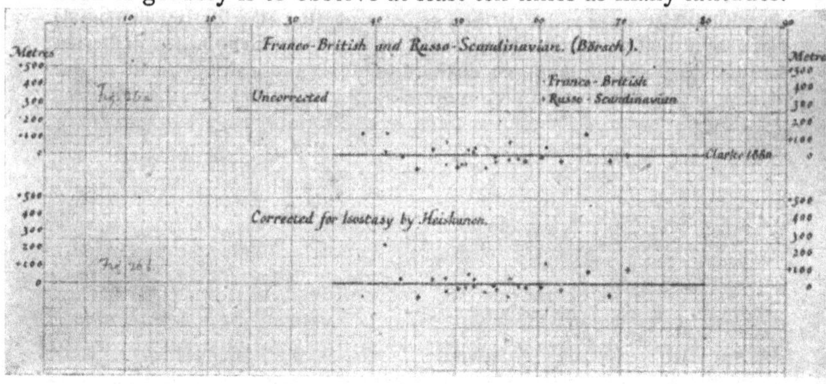

Fig. 26. *Combined Franco-British and Russo-Scandinavian Arcs compared with Clarke's Figure of* 1880
(*a*) *from the Deflections calculated by Börsch,*
(*b*) *from the Deflections corrected for Isostasy by Heiskanen.*

We may conclude by taking the following from the same paper:

The graphical discussion of gravity surveys.

It is natural to examine whether this possibly fruitful and very simple way of looking at the problem can be extended to the discussion of gravity surveys. Happily there is in the memoir of Dr Heiskanen (*Untersuchungen über Schwerkraft und Isostasie*, Veröffentlichungen des Finnischen Geodätischen Institutes No. 4, Helsinki, 1924) a rich store of material ready to hand.

Dr Heiskanen refers his calculations to the expression used by Borrass, which is conveniently designated Helmert I

$$\gamma_0 = 978{\cdot}030\,(1 + 0{\cdot}005302 \sin^2 \phi - 0{\cdot}000007 \sin^2 2\phi).$$

A change in the flattening affects the coefficient of $\sin^2 \phi$. Thus the corresponding coefficients for the flattening of Clarke 1880 and of Hayford 1910 (Madrid 1924) are 0·005248 and 0·005289. If then as before we take Clarke 1880 as the base of ordinates, and plot the differences of Madrid and Helmert I from the curves

$$y = 978\,(0{\cdot}000041) \sin^2 \phi$$

and

$$y = 978\,(0{\cdot}000054) \sin^2 \phi$$

we have the following results (Fig. 27):

We see at once that the difference between Clarke 1880 and Madrid 1924 is equivalent to a variation of 0·020 in the value of gravity between the latitudes 40° and 70°: a difference which should be plainly visible in the gravity determinations of Europe alone.

Let us now plot the gravity anomalies (referred to Helmert I) given in column 2 of Heiskanen's Table XIII, pp. 82 et seqq. for stations in Europe and the Caucasus, corrected for Hayfordian isostasy with depth of compensation 113·7 km. His figures are given as means for each degree square. I have further combined them into means for all the degree squares for a degree of latitude, and have indicated roughly by the size

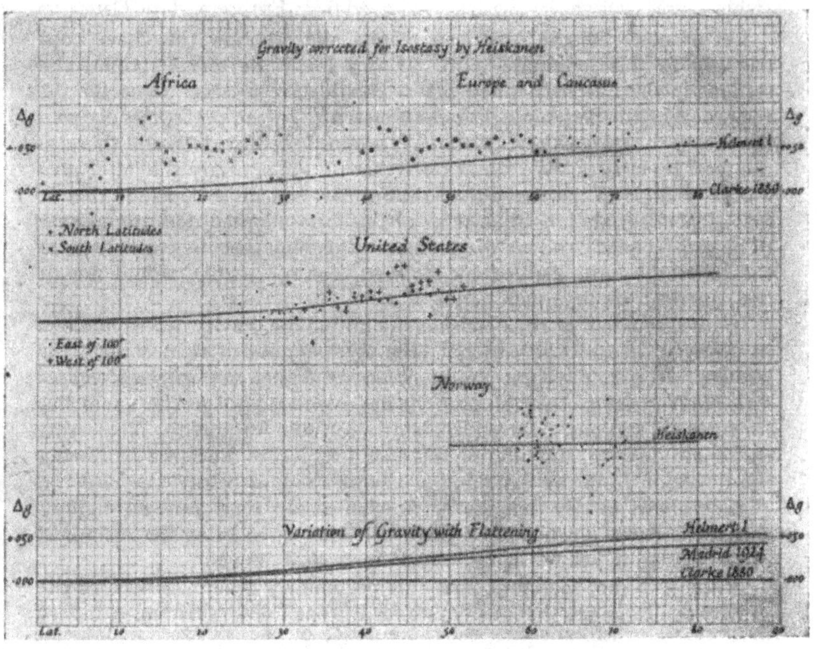

Fig. 27. *Graphical Examination of Gravity Observations.*

of the dots the number of squares thus combined. I have extended the diagram into lower latitudes by adding the stations on the coast of Africa, showing them by crosses instead of dots.

One would say at once that the best curve drawn through this assemblage of points is not far from a horizontal straight line: that is to say, the flattening deduced from these results cannot be far from that of Clarke 1880. In fact, Heiskanen from his solution by least squares deduces from Europe and the Caucasus (the dots) a flattening 293·8, or if Africa is taken in, 294·5. On the other hand, from stations in Asia alone he gets 299·8, and from the United States alone 299·2, while from a combined solution for all he derives 297·4, or very nearly the figure of Madrid 1924. He concludes however that these differences are too large to be explained otherwise than by a longitude term, implying that the equator is elliptical, and deduces finally the following expression for gravity:

$$\gamma_0 = 978{\cdot}052\,[1 + 0{\cdot}005285 \sin^2 \phi - 0{\cdot}000007 \sin^2 2\phi + 0{\cdot}000027 \cos^2 \phi \cos 2\,(\lambda - 18^\circ)].$$

If in this we put $\lambda = 18^\circ$, which is about the mean longitude of the European stations, the expression reduces to

$$\gamma_0 = 978{\cdot}052\,[1{\cdot}000027 + 0{\cdot}005258 \sin^2 \phi - 0{\cdot}000007 \sin^2 2\phi]$$

which agrees very nearly with that derived from Europe and the Caucasus.

Let us now plot the American results and see how they have contributed to this longitude term. If this longitude term is right, the stations in the Eastern United States should give a larger value for the reciprocal of the flattening than the western.* In the preceding diagram the means for stations east of 100° W. are indicated by dots, and those to the west by crosses. We see at once that on the contrary the latter give a greater slope to the curve than the former, and must conclude that at first sight a simple longitude term of the form adopted does not appear to be satisfactory. And we may remark in passing that the coefficients of the longitude term derived by Helmert, cos 2 (λ + 17°), would not fit the American results much better.

In his discussion of 1924 Heiskanen omitted the gravity determinations in Norway, because he was not able at that time to reduce them for isostasy. In a more recent paper: *Schwerkraft und Isostatische Kompensation in Norwegen*, Helsinki, 1926, he has calculated the gravity anomalies for forty-six stations, referred to his own formula for gravity. Neglecting the longitude term, and plotting them with respect to the curve expressing the deviation from the flattening of Clarke 1880, as before, the result in Fig. 27 shows clearly that the Norwegian results tend quite strongly to thrust the curve downwards in high latitudes, or to bring the reciprocal of the flattening below Clarke's value, instead of above it.

It seems to me, therefore, that this preliminary reconnaissance of the gravity results shows that at any rate for Europe Clarke's flattening is very near to the truth, and that the Conference of Madrid was unfortunate in imposing upon Europe a figure of the Earth which may fit the United States tolerably well, though not without suspicious signs of abnormality, but which does not clearly fit Europe at all well, whether one judges by deviations of the vertical or by anomalies of gravity.

The British triangulation.

A word should be said on the state of the British triangulation. It is the earliest in date of any of the great triangulations, and on that account is necessarily somewhat old-fashioned. In a report

* Between longitudes 70° and 100° W. the average coefficient of $\sin^2 \phi$ is about 0·005308; between 100° and 130° W. it is 0·005287, by Heiskanen's formula.

on the triangulations of Europe, made to the International Geodetic Association by General Ferrero, the triangular error of the British work is given as about 3″, with the inference that the work is not of sufficient refinement to be attached to the general European net. This criticism is not quite just. The triangulation of England covered the whole country, and much of it was difficult because the country was so flat. Many of the more unfavourable triangles entered scarcely at all into the arcs of meridian and of longitude and it is hardly fair to burden these with the error of triangulation in other parts of the country.

About 1910 the Lossiemouth base was measured in Scotland by the Ordnance Survey, and connected with the Principal Triangulation. The discussion by Sir Charles Close showed conclusively that the additional strength given to the triangulation by the multitude of figures fully made up for the errors of measurement in individual triangles, and that it is not necessary to contemplate re-observing the whole.

The question arises often, in what circumstances is it worth while to undertake the expense of the greater refinement which shall make a triangulation suitable merely for the framework of the topography into a first class triangulation fit to take its part in the general problem of determining the size and shape of the World? The answer must depend to a great extent upon the relation of the country in question to the principal land masses of the World. If it is isolated, and of moderate size only, then it can contribute little to the solution of the problem. But wherever there is a possibility of junction with other extensive work of the same kind it should be a point of honour with the country to contribute its share in the solution of the great problem.

Geodetic levelling.

The immediate practical importance of the main lines of precise or geodetic levelling is to provide a foundation for all the subsidiary lines upon which the local lines of levels, and in Great Britain the contours also, are based. Its ultimate scientific, and perhaps also practical importance, is to discover whether the whole country is gradually rising from the sea, or sinking, or tilting; and at what rate the mountains are growing or becoming denuded.

Within the limits of this book we cannot deal with the instrumental precautions which are essential in the conduct of precise levelling. The general procedure is quite similar to that which we have sketched in Chapter IX, page 202; but there are many precautions to be taken to avoid the small but cumulative effects of temperature, of refraction, and of local deviations of the vertical.

It is now realised that the original net of levelling in Great Britain was not observed with sufficient precautions to give an ultimate verdict on the above points. In particular the determination of mean sea level was not satisfactory. A revision of the principal levelling has been completed recently, great attention has been paid to the important question of placing the fundamental bench marks upon solid rock, the lines carefully chosen to avoid districts where subsidence might be caused by mining, and three tide gauges established at Newlyn, Felixstowe, and Dunbar.

In India the great earthquake of 1905 provoked a very interesting enquiry into the changes in relative heights produced by the shock. It was found that the difference of height between Dehra Dun and Mussooree had been diminished 5·5 inches, and it was at first supposed that Mussooree had subsided by this amount. But a revision of the line of levels into the plains showed that this was not the case. Mussooree in the Himalayas and Saharanpur in the plains were found to be at the same relative height after the earthquake as before; but the intermediate station of Dehra Dun was higher by five inches.

There is great reason to think that this movement was only an exaggeration of a movement that is always going on, and that the Himalayas are gradually pressing forward and up the lower lying ranges to the immediate south of them. For this reason several lines of precise levelling have been carried up into the Himalayas, and connected with the older formations to the south. It is hoped that in time this work may give information on the rate of growth of the mountains.

Similar enquiries undertaken by the Japanese Surveys after the great Tokyo earthquake of 1923 have given much detailed information on the larger changes of level, up to 2½ metres, which came with that shock. (See *Geog. Jour.* October 1927.)

Chapter XII

PHOTOGRAPHIC SURVEYING

Rough surveys from Panoram photographs.

A PANORAM camera covers a field of about 130°, and a series of three panorams carefully taken to overlap, so that they cover a complete round of angles from the station, may be used to good effect. The three prints are fitted together and the angular scale value determined from the distance between corresponding points at the beginning and end of the strip, whence with a divided scale the relative bearings of all conspicuous features may be measured to an accuracy of a few minutes of arc. A fairly correct map of the Mount Everest region was made by this method before any of the survey results were received in England (*Geog. Jour.* February 1922).

The "Canadian" method.

This elementary method of using photographs in survey was first developed in France by Laussedat, but since its modern use has been principally in Canada it is often referred to as the Canadian method. Its principles are very simple, but its application rather tedious.

The photographs are taken with a camera which can be levelled like a theodolite so that the plate is strictly vertical, when it is brought up from its plate holder against stops in the camera which define the focal plane, and in which V's are cut to define the horizontal and vertical lines through the foot of the perpendicular to the plate from the optical centre of the lens. By preliminary triangulation or resection with the theodolite, which in the Canadian method is separate from the camera, the position of the camera station and the azimuth of one conspicuous point in the field are determined. From the negative a print of known enlargement is made on paper; lines are drawn from all the points to be determined to the horizontal line through the centre; and the feet of these perpendiculars, with the centre, and the perpendicular from the azimuth mark, are marked off on a slip of paper.

The camera station is plotted on the map and the azimuth laid off. The marked strip is then laid on the map at a distance from the station point equal to the focal length multiplied by the enlargement, and in such a position that the perpendicular from the station to the strip cuts it at the centre mark, while the azimuth line cuts it at the azimuth mark. Rays drawn from the station mark to the other points on the slip then evidently pass through the map positions of those points. The process is repeated at a second station, and the intersections of pairs of corresponding rays give the positions of the required points on the map. The difference of height between any such point and the camera station is found by measuring on the enlargement the length of the perpendicular to the horizontal line, which is to the distance from the station along the ray to the strip as the actual difference of height is to the distance in feet scaled off from the map. The calculation is simplified by various mechanical devices.

The process has the merit of simplicity, but the defect that there are often not many points which can be certainly identified on the pair of photographs, without the use of some form of stereo-comparator. When a pair of plates is combined stereoscopically the identification becomes obvious, and as many points as may be desired can be plotted with little trouble. The attention of the inventive may be directed to the need for a stereoscopic plotter simpler, even if much less accurate, than the instruments described in Chapter XIII. The map of Mount Everest and the Chomolungma massif was made by Major Wheeler, of the Survey of India, by the Canadian method. (Special publication of the Survey of India, in three sheets.) The map included in the book of the Third Mount Everest expedition is based upon this, but with rock and glacier drawing by the late M. Charles Jacot-Guillarmod.

Stereographic survey.

In the opinion of the author the future undoubtedly lies with stereographic survey. No one who has spent time searching for pairs of corresponding points on separate photographs, and has then seen how stereoscopic combination makes the identification automatic and absolute, is likely to be content for long with the older method. The first stereoscopic surveys were perhaps those made in Austria with the Stereo-comparator of Pulfrich. This involved plotting point by point after simple calculation. About the same time Captain Vivian Thompson of the Royal Engineers devised the first elementary form of an automatic plotter (*Geog. Jour.* May 1908) with which a trial survey was made about Keswick, some work was done in Fiji, and Lieut. Kenneth Mason made a map of the Pamirs which has remained unpublished. For one reason or another Thompson's instrument never received the mechanical development which it deserved, and the next step was taken by Lieut. von Orel of the Austrian Military Geographical Institute, whose Stereo-autograph was perfected by the firm of Zeiss. This was the first instrument for automatic plotting of contours that passed the experimental stage and reached commercial success; but its use was impeded by grant of sole concessions, and the requirement of royalties on gross receipts. It could never, on those terms, come into scientific use. The concessionaires for France, the *Société française de Stéréotopographie*, have done much work on large scales for railways and mining companies, but it does not seem that it has ever been employed on geographical scales. The stereographic method of survey is not easily applied to flat country without commanding points of view; and in hilly country there is dead ground that can be covered only by great multiplication of photographs. The French company found by experience that about three-quarters of the ground can be surveyed economically by the direct stereoscopic survey from ground stations; but that the dead ground can be completed very readily with the plane table, when all the detail round the edges of the dead ground is fixed from the photographs.

Recent developments in Photostereographic surveying.

The principle of the stereoautograph does not seem to lend itself to development in three-dimensional survey, with the axes of the plates inclined upwards to take in the heights of mountains or plunging into deep valleys from ground stations, and still less to dealing with air photographs either vertical or oblique. The machines which have been developed during the last few years all utilise in different ways the principle of the *Bildmesstheodolit*, otherwise called the Photogoniometer, in which the plates are viewed through objectives identical with those which took them. The pencils of rays which were brought together to make the images in the cameras are thus effectively reconstituted, and the geometrical solution is made to depend on angles and not on the linear projections upon the plates.

The positions in space of the plates at the moment of exposure are in effect reconstituted; the plates are viewed in a stereoscope carrying a "floating mark"; and the movements which bring the mark into apparent contact with detail on the photograph bring the pencil into position to plot the detail in its place on the horizontal plan, and also determine its height; while if the height mechanism is clamped at a given value the mark can be made to traverse and the pencil to draw the corresponding contour.

There has been great activity in developing machines to perform this complicated feat, and many claims have been made for them which are hard to substantiate. We may divide the machines first into two classes: those in which the rays from the photogoniometers are brought directly into the stereoscope; and those in which they are projected on a screen and the projected images thus viewed. In the first class are the Autograph of Heinrich Wild, the Autocartograph and the Aerotopograph of Hugershoff, the Stereoplanigraph of Zeiss, the Fourcade Stereogoniometer of the British War Office, and the somewhat similar machine invented by Poivilliers and in use by the Service Géographique de l'Armée. In the second class are the Nistri Photocartograph and the Barr-Stroud Photogrammetric Plotter made for the War Office and familiarly known as Big Bertha. Machines of the second class are, in the opinion of the writer, optically inferior to the first and may be omitted in a brief review.

The Autograph of Wild was developed first for stereo-survey from ground stations, and later adapted for air survey. The others in this class have been designed more particularly for air survey, and will be considered later. The Autograph emerged from its experimental stage, necessarily long and costly, and came into regular use in Switzerland for work on large scales about 1924. In his Vermessungsbureau at Flums Dr Helbling soon had two machines in constant use for railway surveys in Asia Minor and the Andes; and the Swiss Federal Survey Department found that the contours drawn by the machine in mountainous country are so much more accurate than those made by the older methods that they undertook a re-survey of their mountains.

These facts are sufficient answer to critics who demand the cost per square mile of the process, as compared with the older methods. A national survey finds it worth while to re-survey by the new method, though the existing maps are of what used to be considered high quality; and a private enterprise builds up a lucrative business.

Moreover, there are occasions in which cost is a minor consideration, and the possibility of doing the work at all the chief concern. When conditions are difficult, and an expedition only a short while in new country—or when it is a question of obtaining as much information as possible without entering a country, as in much military reconnaissance—then the method may be deemed invaluable.

The stereographic Survey of the Shaksgam.

The first attempt to survey stereographically on small scales with the most recent instruments was made by Major Kenneth Mason on the Shaksgam Expedition of 1926. Pairs of plates taken with the Wild Photo-theodolite lent by the R.G.S. were measured on the Wild Autograph in 1927 on the scales of 1/50,000, 1/100,000, and 1/250,000. The results surpassed all expectations; it was found possible to contour peaks up to thirty and forty miles away in the clear air of the Himalaya. There was of course much dead ground; but as a *tour de force* it was a great success, and justifies the belief that there is a great future for this type of survey in difficult country. See the paper by Major Mason in the *Geog. Jour.* LXX, 342, Oct. 1927.

There is much difference of opinion on the relative value and future importance of the different types of photographic survey. One school of thought favours the "Canadian Method" for all ground work, but admits the necessity of stereographic methods for air survey; a second favours the general use of air photographs, but is content to work them up by the relatively simple but laborious graphical processes; a third school goes all-out for stereoplotting; a fourth, to which the author inclines, believes that stereographic survey from ground stations will play a considerable if not a principal part in the future, supplemented by air survey for the dead ground, tied to points already determined by the ground photographs. Methods must of course vary with the ground, the accuracy required, and the money available, and while methods are still experimental one cannot be dogmatic. One may believe that in the near future it will be the rule to take a round of plates with a photo-theodolite at every mountain trig. station, as a matter of routine, and to determine from them a good part of the topography, with such a sufficiency of points round the margins of the dead ground that the map may be completed by plane table with great speed and accuracy. Or the dead ground may be covered by air survey. But in the opinion of the author there is some danger that too much attention will be paid to the difficult case of air photographs, which have the grave inconvenience that they are taken from points that have to be determined implicitly if not explicitly with the height, bearing, and tilt, before anything more can be done; whereas plates taken at ground stations fixed trigonometrically start with the initial advantage that the position of the base is known with accuracy far greater than can be derived from the photographs themselves; and in searching for cause of discordance one principal trouble is eliminated.

Recent experience suggests that for extensive, systematic, rapid, and thoroughly organised surveys the future lies with the air photographs, but that we must not neglect the great value of ground stereo-survey in exploration, as in the Shaksgam, or in the study of special problems which may more particularly interest the geographer, such as the growth and movements of glaciers, the changes in deltas and screes, the characteristics of land-forms, the structure and the denudation of volcanoes, and many such matters.

Mapping from air photographs.

Mapping from air photographs developed during the Great War in two directions: the reconnaissance of unmapped country, as east of the Jordan, with no fixed points, or an occasional doubtful position; and the revision of old or mapping of new detail when a good framework of road junctions, railway crossings, etc. is already in existence. Since the war progress in the first direction has been largely in learning how to fly straight and level, so that a series of overlapping photographs may be little distorted by tilt, preserve a fairly uniform scale, and cover the ground in strips without gaps. These conditions are hard to fulfil, even in peace, but much has been done by fitting gyroscopic control to the aeroplanes; and we were in 1931 just reaching the stage when a strip of photographs with 60 per cent. overlap could be adjusted to an initial pair of fixed points and closed on another pair perhaps ten miles away. For air photographs reduced individually one must have four ground points per plate. Simple methods have been developed for fitting the photograph to these points, with fair success when the ground may be taken as practically flat. But when this is not so the difficulties are much greater. For mapping hilly country stereoscopic methods are essential; but in the present state of the subject, still experimental, it is impossible to consider any method as standard.

Air photographs are distorted to a small extent by a progressive lag due to the focal plane shutter; by swing of the aeroplane during the exposure; and occasionally by the distortion of the lens or of the film during development, though these last two should never be serious. Photographs are gravely distorted by tilt. It is fairly easy to fly straight, except under fire, but a very slight defect in the rigging of the aeroplane makes the machine gradually gain or lose height when the pilot thinks that he is flying level; hence the photographs may continually alter in scale, even when the tilt is not serious.

A photograph distorted by tilt may be adjusted whenever four known points well distributed on the ground covered by the plate can be identified with the corresponding images on the plates. The following methods have been employed:

1. Transfer of points one by one from photograph to map by interpolation with proportional compasses. This is tedious, and inaccurate, because the scale of the distorted photograph is not uniform along a straight line.

2. Geometrical construction of points on the map, one by one, by methods depending on the properties of anharmonic ratios. This is accurate but tedious, and cannot be used for many points, though on occasion it is valuable. See page 248.

3. Rectification by camera lucida, adjusted so that the four known points of the photograph are projected on the four points plotted on the map, when any other detail can be traced by the pencil. This requires an elaborate installation of the camera lucida, with full adjustments for change of scale and tilt. The adjustment requires a thorough knowledge of the properties of perspective projection, and the work of tracing is very trying to the eyes.

4. Rectification by re-photographing in a special camera with lens giving a very large flat field, and with full adjustments for tilt of original

negative and of plate. This method may serve well in flat country, and several instruments have been devised for the process, of which we shall give the geometry later. See page 250.

5. Rectification by constructing a perspective diagram, to transfer the map grid to the photograph, when the detail may be sketched from the photograph square by square. This method has been much used in Canada for surveying flat lake country. It is of no use in hilly country, and not much good in flat country unless the detail and the horizon are very clearly defined. See page 253.

6. Plotting stereoscopically with a simple stereoscope. This has been developed by the Air Survey Committee. It required extensive ground control both in position and height, in its original form of the "Arundel method", but is now developed into a rapid and effective approximate method. See page 255.

7. Plotting and contouring automatically with elaborate stereo-plotting machines. The machines have been brought to a high pitch of optical and mechanical perfection: the difficulty is still with finding the position and tilt of the camera at the two stations. See page 256.

8. And finally, giving up any attempt at rectification, trying to fly as level and straight as possible, assuming the photographs vertical, and patching them together into a mosaic as well as may be.

To transfer a single point from photograph to map by the "four point" method.

Find four points *a*, *b*, *c*, *d* on the photograph corresponding to four points *A*, *B*, *C*, *D* on the map. Join *a* to *b*, *c*, *d* and *A* to *B*, *C*, *D*. Join also *a* to *p*, the point to be transferred. Lay a strip of paper across the pencil of lines from *a*, and mark where they cut the strip. Place the strip on the pencil from *A*, and move it about until the three ticks corresponding to *ab*, *ac*, *ad* fall on *AB*, *AC*, *AD*. The line through *A* to the tick corresponding to *ap* will then pass through *P*, by a well-known property of perspective.

Repeat the process using *b* and *B* as vertices. The intersection of *AP* and *BP* will give the required point *P* on the map corresponding to *p* on the photograph. It is tedious to make a whole map like this, but the process is useful in constructing single points.

For example, given four known points on a photograph which is tilted, to find the horizontal through the principal point: that is, the point where the perpendicular through the optical centre (or more precisely, the second principal point) of the camera objective cuts the plate. As before, let *a*, *b*, *c*, *d* be four points of the photograph corresponding to four points *A*, *B*, *C*, *D* of the map. Draw the parallelogram *ABED*, and by the four point method transfer *E* to *e* on the photograph. If the third diagonal of the quadrilateral *abed* can be constructed within the limits of the photograph, this is the trace on the photograph of the true horizon, since pairs of opposite sides of the quadrilateral intersect upon it in the two "vanishing points". And the parallel line through the principal point is the "axis of tilt".

But often the horizon trace lies far away outside the limits of the photograph. In this case draw a transversal *g'fh'* through the intersection of the diagonals, cutting *ad* in *g'*, and make *fh'* equal *fg'*. Draw *h'h* parallel to *ga* to cut *eb* in *h*. Then *hfg* is a horizontal. When the photograph is nearly vertical *be* is nearly parallel to *ag*, and the method fails, as is natural.

If θ is the inclination of the optical axis to the vertical, or the tilt, then the distance of the true horizon from the principal point is the focal length $\times$ cot θ, and the tilt may be calculated; but in practice it rarely happens that four known points can be identified upon a photograph with so great a tilt that the true horizon comes within the limits of the plate. And there is the further serious restriction that the above theory applies only when the ground is flat.

The rectification of air photographs.

The rectification of a tilted air photograph is its transformation to what it would have been if the axis of the camera had been vertical at the moment of exposure. It is difficult to secure that a photograph intended to be vertical shall not have a tilt of a degree or two, while photographs are often taken deliberately at large tilts to cover more ground, or ground that cannot be flown over. Hence practically most air photographs require rectification. The point-to-point method described above is an elementary method of rectification; but something more speedy is required.

We may assume that in future the focal length of the lens will be known accurately, and the principal point of the photograph will be defined by accurate fiducial marks on the margins. When focal length and principal point are known, and the ground is flat, the angle and direction of tilt and the height of the lens above the ground can be found from three known points by the construction known as Finsterwalder's method, described in *Graphical Methods of plotting from Air Photographs*, page 34. When the ground is not flat, they may be found by the more elaborate constructions of pages 59 et seq., or they may be calculated by somewhat long and intricate trigonometry. Alternatively the tilt may be found by a "tilt-finder", which is a rather complicated machine constructed to the designs of the Air Survey Committee, in which the three known points are set by their co-ordinates and the photograph adjusted to fit them in perspective. These methods are all still in the experimental stage.

But supposing that the tilt is known, and the height, then the photograph may be rectified photographically, by a method whose theory is simple and interesting, but not usually found in the text-books of geometrical optics.

Theory of photographic rectification of air photographs distorted by tilt.

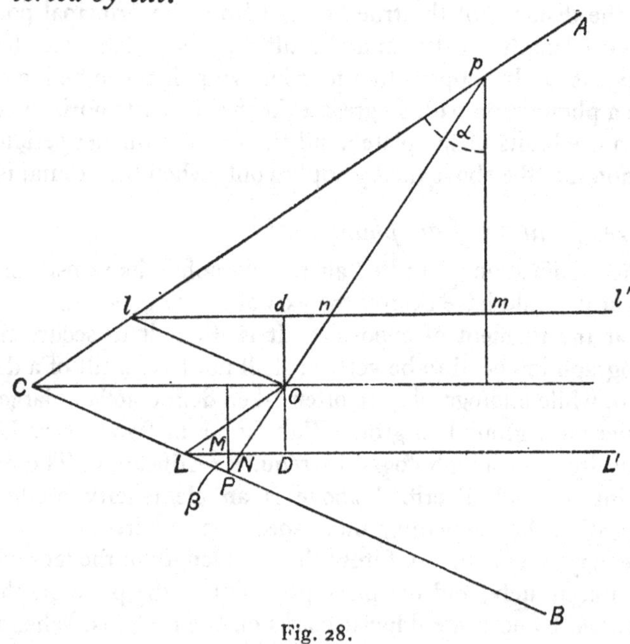

Fig. 28.

I. Optical Theorem: Condition that two lines *CA* and *CB* shall be optically conjugate for a lens with centre *O* and axis perpendicular to *CO*.

Draw *OL*, *Ol* parallel to *AC* and *BC*,

ll', *LL'* parallel to *CO*,

pnONP making any angle with *CO*,

pm, *dOD*, *PM* perpendicular to *CO*.

Then $$\frac{pm.PM}{Od.OD}=\frac{pn.PN}{On.ON}=\frac{pl.PL}{Cl.CL}=\frac{pl.PL}{OL.Ol}=1,$$

since the triangles plO, OLP are similar.

But $Od = OD$. Hence $pm.PM = OD^2$ and p, P are conjugate points for a lens at O of focal length OD, the lines ll', LL' being in the principal focal planes, and the lens being supposed to have a flat field of indefinite extent.

And corresponding points in lines through P, p perpendicular to the plane of the figure are also conjugate.

Hence two planes through AC, BC, perpendicular to the plane of the figure, are completely conjugate, and the lens will form a sharp image of one on the other, though they are not parallel.

This elegant theorem seems to be due to a Captain Scheimpflug of the Austrian army.

If the planes CA, CB make angles α and β with the axis of the objective, $CO = OD (\tan \alpha + \tan \beta)$.

II. Perspective Theorem.

If the photograph is taken at height h with a lens of focal length f whose axis is inclined to the vertical at an angle θ, the scale of the photograph, that is, the scale of the horizontal line through the centre of the plate, is $f \cos \theta / h$.

We wish to rectify this inclined photograph, so that it fits a map of scale s, by re-photographing it with a lens of focal length g.

Let O_1 be the camera objective,

CA the plane of the plate,

O_1c perpendicular to CA, and hence c the plate centre,

$O_1c = f$ the focal length of the aeroplane camera,

and O_1H parallel to the ground when the photograph was taken.

Draw CG also parallel to the ground (and to O_1H) at distance from it $= sh$, cutting AcH in C.

The projection of the ground on CG, from centre O_1, is the map on scale s which is required.

Draw a line through C at perpendicular distance g from H to cut in O_2 the circle described about H with radius HO_1, and draw CB parallel to HO_2.

Then by the optical theorem above, planes through CA, CB perpendicular to the plane of the figure are conjugate for a lens at O_2 of focal length g, and axis perpendicular to CO_2, in the plane of the figure.

Also, by a well-known theorem of perspective projection, the projection of the plane CA on CB with centre O_2 is identical with its projection on CG with centre O_1, since the centres O_1, O_2 are equidistant from H.

Hence we have a method of re-photographing the air photograph in plane CA to make it fit the map on any desired scale.

To set up the apparatus we must know g, the focal length of the copying lens, and calculate the angles α and β, and the distances O_2D, O_2E, and cD. Our data are f the focal length of the aeroplane camera, θ its inclination to the vertical, h its height above the ground, and s the scale of the map desired. Of these θ and h will be known only approximately. But sup-

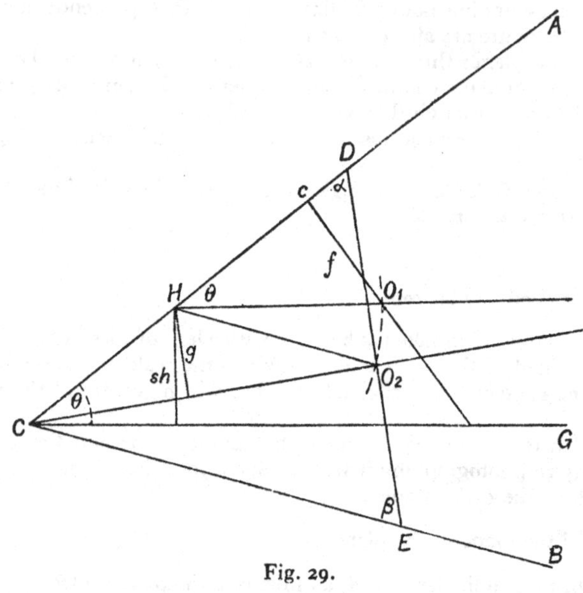

Fig. 29.

posing for the moment that they are sufficiently determined: then it is easy to deduce from the figure that

$$\cos\alpha = g\sin\theta/sh,$$

$$\cos\beta = g\sin\theta/f,$$

$$O_2D = g\,(1 + \tan\beta\cot\alpha),$$

$$O_2E = g\,(1 + \tan\alpha\cot\beta),$$

$$cD = f\operatorname{cosec}\theta\,(\sin\beta\operatorname{cosec}\alpha - \cos\theta).$$

The difficulty in the application of this theory is that θ and h cannot at present be determined with sufficient accuracy by any observations made when the photograph is taken. Various methods have been proposed for finding θ from comparison of four or three known points on the photograph and the map, as briefly indicated above: but it cannot be said that any process yet devised is really satisfactory. If θ is found, the process will give also the horizontal diameter of the photograph, whence it is simple to find its approximate scale r in that direction, and since $h = f \cos \theta / r$ to find α from $\cos \alpha = gr \tan \theta / fs$. Then setting up the apparatus and adjusting β to get good focus, we shall have an approximate fit of the distorted photograph to points plotted on CB.

There are descriptions of apparatus constructed on this principle, but little information as to its accuracy and convenience. The whole subject offers a promising field for research.

A photograph so rectified does not correspond to a photograph taken with the camera axis vertical unless the ground is absolutely flat. Consider an oblique photograph of ground with a house in the middle distance: one or two sides of the house will appear in the photograph, and will appear also, though altered in shape, in the rectified photograph. Thus the two ends of a vertical line on a wall, which should appear as one point on the map, will be two on the rectified photograph. It is easy to show that the line joining any such pair of points passes through the point on the photograph corresponding to the "plumb-point" on the ground: that is, to the point vertically under the lens centre at the moment of exposure. Hence if the plate plumb-point can be determined on the rectified plate, we have a true direction for each point; and if two photographs taken more or less at right angles be used, the intersection of the two directions will give us the true position of the point projected on the map plane, while the displacements of the point from the intersected position will give at least an indication of its height above that plane. This principle is the foundation of the method of stereoscopic plotting from approximately vertical photographs in No. 6 above.

Rectification by perspective grids.

Since plotting point by point is impossibly tedious, and photographic rectification not yet very practicable for large tilts, we may seek a method of transferring a rectangular grid on the ground or the map to the oblique photograph, and sketching in on the map the detail of the photograph square by square. If we have three fixed points recognisable on the photograph, and know the focal length and the principal point, we can determine the height and the tilt, and thence by methods of perspective geometry which are practicable, if complicated, we may draw the corresponding perspective grid upon the photograph. If the photograph is so oblique that the horizon comes upon the plate we can do with only two fixed points. But even so, the construction of a perspective grid for each plate is a long business (see *Graphical Methods of Plotting*, chap. IX, p. 70), and in Canada, where the method has been practised with most success, in the survey of the lake country which is essentially flat, they have found it simpler to make a whole set of grids corresponding to a graded series of heights and tilts within which the photographs are taken, and to select the grid which is found to fit the photograph most nearly.

With an ascertained or assumed height h and tilt θ we may construct a perspective grid thus:

Suppose that a plate 6 inches square is taken with a lens of 10 inches focal length, at a height of 7000 feet and at a tilt of the optical axis 50° from the vertical; the map is to be compiled on the scale 1/10,000, with a grid of 1000 feet on the ground or 1·2 inches on the map. Let p be the principal point, ph the principal line, or line of greatest slope, and pl the horizontal line through p, or the axis of tilt. Then, as we shall see later, the projection of the true horizon on the plate is parallel to pl, and at a distance

$$ph = f \cot 50^\circ = 8{\cdot}39 \text{ inches from } p.$$

The projection of the grid on the plate divides pl into parts each

$$f \cos 50^\circ \times 1{\cdot}2 \times 10{,}000/7000 \times 12 = 0{\cdot}918 \text{ inch.}$$

Divide the line pl thus and join the dividing points to h on the projected horizon. The resulting lines are the projections on the plate of one set of parallel grid lines.

Lay off a distance $hk = f \operatorname{cosec} 50^\circ = 13{\cdot}05$ inches along the horizon, and join pk. This is the projection of a diagonal of the grid: and the other set of grid lines on the plate is formed by parallels to the horizon trace through the intersections of this diagonal with the first set of grid lines.

When the perspective grid is thus drawn on the plate, the detail can be sketched square by square on to the map compilation, provided that the ground is flat. The geometry is simple: in practice the horizon line and the intersections of the diagonals with it will often be at distances too great for plotting, in which case the grid must be calculated from the theory which follows.

This is all on the assumptions (1) that the ground is flat, as in a river delta, or on the lake plateau of Canada; and (2) that the height and tilt of the camera are known. The various methods of calculating these quantities from the co-ordinates on the plate of these known points, and the focal length, are so long that they are little used, being superseded by experimental methods such as that of the tilt-finder. Good progress has been made in constructing such machines, but they have hardly yet arrived at perfection; and it is by no means certain that this method of perspective diagram plotting has any future as a general method of air survey.

In country which is very nearly flat, and in a climate which allows a clear image of the horizon on a much tilted photograph, this Canadian method is practised with considerable success. For a standard camera sets of grids are constructed on celluloid for ranges of altitudes differing by 25 feet, and for differences of tilt which shift the apparent horizon by steps of 0·1 inch on the plate —the distance between the apparent and the true horizon (refraction and dip) being calculated (*Graphical Methods*, pp. 57–59). Any actual photograph is fitted to a grid which has the proper horizon distance from the edge of the plate, and is made for the height shown approximately by aneroid. The distance between two control points thus rectified, compared with the true distance, gives a correction to the assumed height, and designates that grid in the prepared set which fits the case most nearly. The detail can then be sketched square by square. It should be noted that when the horizon is visible only two control points are required; and that in practice these control points are selected from the photographs, taken in strips with overlap, and are fixed on the ground by a traverse, before the photographs can be plotted. The resulting plots are on grids of varying orientation, which must be fitted to the standard grid by the control points, and the detail transferred. Alternatively, one may start with the photographed positions of the control points, the principal point, and the horizon, and construct for any individual plate the perspective of the orientated map grid (*Graphical Methods*, pp. 70 et seq.) but this takes longer than the method described above, though it is presumably more accurate.

Plotting with the Barr-Stroud field stereoscope.

This method was developed by the Air Survey Committee, and the first results, for a district near Arundel, are published in their *Professional Papers*, No. 3 (1927). The photographs, approximately vertical, were taken in strips, with a longitudinal overlap of about

60 per cent., and a smaller but considerable lateral overlap. The first step is to transfer the principal point of each photograph to the preceding and following, by the use of stereoscopic markers explained *loc. cit.* p. 8. This provides a series of "principal point bases", which would be equal and collinear if the flying and photography were perfect. From these minor control points are fixed by intersections, on the assumption that directions from the principal point are true, which is correct only if there is no tilt. At the same time the ruling points for contouring were selected in the stereoscope and marked on a series of prints, sent into the field with a party who tied the minor control to the principal triangulation by theodolite, and determined the heights of the ruling points by traverses with a battery of aneroids. With the minor control thus adjusted the plates were finally plotted, and contoured by sketching with the stereoscope. The first results were promising: but it should be observed that there is a good deal of difference between this process, involving much work on the ground, and the popular idea of air survey.

The results of a second experiment, made in more mountainous country, are published in *Professional Papers*, No. 6 (1930). Since then the method has been used in Transjordan, the Suez Canal area, and on the boundary survey with Italian Somaliland. The latest developments are described in Prof. Paper No. 8 (1933) by Lieut. J. S. A. Salt, R.E.

Plotting with stereo-plotting machines.

The machines have been brought to great perfection: thus the Autograph (described on page 275) can deal with pairs of photographs taken at tilts up to 40° from the vertical or the horizontal. But before these can be set in the machine for automatic plotting and contouring, it is essential that the camera positions should be known within a metre or two. Their determination by measurement and calculation having proved almost impossibly long, it has been found possible to use the stereo-plotters as elaborate tilt-finders: but it is not yet possible to say that the method has reached its final form: whereas stereo-survey from ground stations is well developed. At first in Swiss practice the latter were called upon to provide the ground control for the air photographs, which were employed only to fill in the dead ground that could not be dealt with from ground stations. Latterly the air photographs have played a larger part. It should be noted that the Autograph is well adapted to deal with a single pair of photographs; but is not designed for working up a long strip of overlapping plates, dealing first with Nos. 1 and 2; then removing 1 and adjusting 3 to 2; and so on. For this, which seems likely to become the standard method of regular air-survey, the Stereoplanigraph in its latest form, the Fourcade Stereogoniometer, and the Poivilliers machine, are especially designed.

The geometry of a tilted air photograph.

We may hope that without tedious definition the geometry is almost self-evident in the accompanying figure. *ABCD* is a rectangle in the horizontal ground plane and *CDEF* the plane of the plate, cutting the ground plane in the "perspective axis" *CD*. The "principal plane", perpendicular to both, contains the centre of projection *O*—the camera lens; the "principal point" *p* of the plate, at the foot of the perpendicular from *O*; the principal point *P* of the ground plane; the "ground plumb-point" *M*, foot of the perpendicular from *O* on the ground plane; the "plate plumb-point" *m*, where this perpendicular cuts the plate plane; and the point *h* where the parallel through *O* to *PMT* cuts the plate plane. The actual plate has a "swing" so that its edges are not parallel to the principal line *mphT*. The angle *pTP* is the tilt θ; *Op* the focal length is *f*; and *OM* the height of the camera lens is *H*.

The ground has a square grid, squared to the principal plane. If we join *h* to the points where one set of lines cuts the perspective axis, and produce backwards, we have the "plate meridians". The line through *h* in the plate plane parallel to the perspective axis is the trace of the true horizon on the plate, and the parallel through *p* is the "axis of tilt". The remaining parallels may be constructed by producing the diagonal *BP* of the ground grid to meet the perspective axis in *K*; the points where *pK* cuts the plate meridians are on the plate parallels, which may be drawn through these points parallel to the horizon trace; or better, by drawing the other diagonal, and joining the intersections in pairs. The graphic construction of such a perspective grid to full scale will generally be impracticable; but the expressions for distances, angles, and co-ordinates are readily derived: it is easily seen that

$$ph = f \cot\theta; \quad Oh = hk = hk' = f \operatorname{cosec}\theta; \quad pm = f \tan\theta;$$

$$hT = H \operatorname{cosec}\theta; \quad MP = H \tan\theta; \quad OP = H \sec\theta.$$

If $nOp = \alpha$, $\quad pn = f \tan\alpha.$

And
$$\frac{PN}{PO} = \frac{\sin\alpha}{\sin(\frac{1}{2}\pi - \theta + \alpha)},$$

whence
$$PN = pn\,.\,H \cos\alpha \sec\theta \sec(\theta - \alpha)/f,$$

and
$$NQ = nq\,.\,H \cos\alpha \sec(\theta - \alpha)/f.$$

Hence taking the axis of tilt and the principal axis as the axes of x and y on the plate, the relations between these and the corresponding X, Y on the ground are given by

$$y = \frac{fY}{H \sec^2\theta - Y \tan\theta}, \quad x = \frac{X(f\cos\theta + y\sin\theta)}{H}.$$

The bearings from the principal axis at the principal points are related by the equation $\tan QPN = \tan qpn . \cos\theta$ and the bearings at the plumb-points by $\tan QMP = \tan qmp . \sec\theta$.

Hence it follows that the axis of tilt is divided by the meridians into parts which are $f \cos\theta/H$ times the corresponding parts on the ground; and that the diagonal may be drawn from $\tan kph = \sec\theta$.

If the ground grid makes an angle with the principal axis, the line through O parallel to the grid to cut the horizontal in h' gives the vanishing point of the plate meridians.

If the ground is not flat, it is evident by considering the plane *PMOpm* that the bearings on the plate from *m* are unchanged; but all other bearings are altered.

Most of the trigonometrical relations which are scattered through the literature of the subject may be derived conveniently from this figure. The practical interest of this whole theory of perspective grids is limited by the difficulty of finding the tilt and the axis of tilt of any photograph. But the same theory underlies the theory of stereo-plotting machines, and it is therefore of permanent interest.

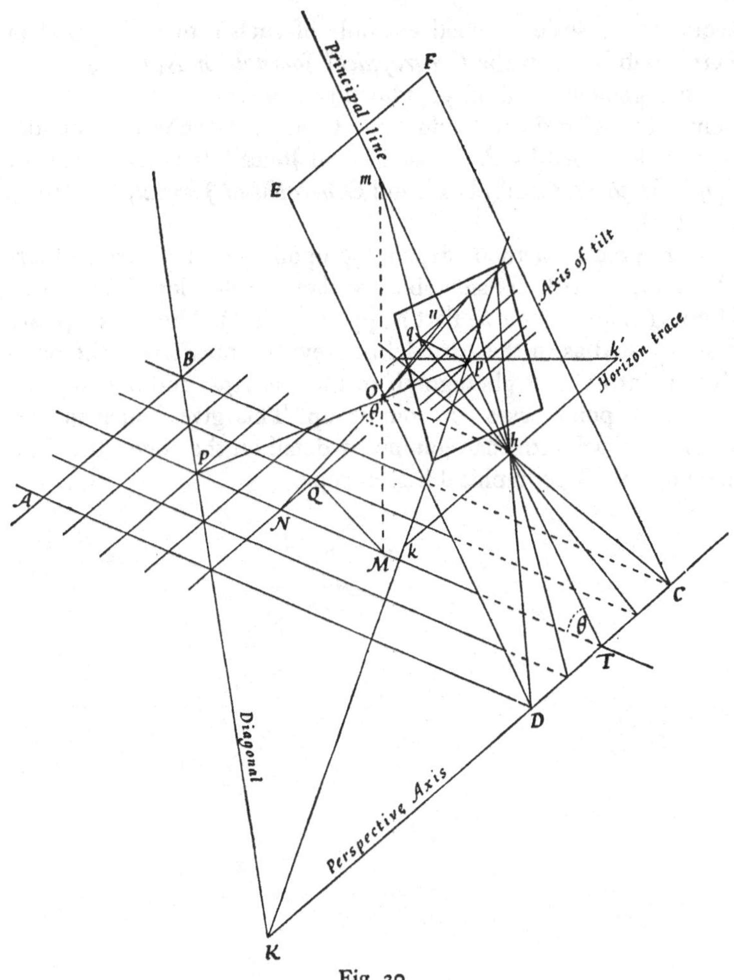

Fig. 30.

Mosaics.

A mosaic is a composite picture obtained by pasting together a number of overlapping vertical air photographs. The common method of doing this by trial and error fit of detail along the margins of the photographs very soon breaks down, because of the rather large distortions produced by the inevitable tilts of a degree or two, and the inevitable small variations of height, and consequently of scale. A good example of such a mosaic is that of Petra, published in the *Geographical Journal* for April 1924.

An ingenious method of adjusting a mosaic to a few control points, by fixing the individual points on an elastic web as foundation, is described by Professor Melvill Jones (*Aerial Surveying by Rapid Methods*, Camb. U.P., and *Geographical Journal*, June 1923, LXI, 413).

More precise methods depending upon controlled strips of air-photographs with considerable overlap are developed by Lieut. Hotine (*Simple Methods*, Part II, pp. 48 et seq.). The key strips are first adjusted as in the Arundel Survey, by transferring the principal point of each photograph to the overlaps, and aligning the "principal point bases" in joining up. This gives much greater accuracy in azimuth than joining by detail on the edge. It is best done by the Topographical Stereoscope.

Chapter XIII

SURVEY INSTRUMENTS

THIS book is not intended to supply the place of a technical handbook in which all the details of the instruments, the methods of adjustment and of observation, are given minutely. Such information may be found in full in the *Textbook of Topographical Surveying*, to which reference is made so often, or in other technical works. But it will be well to give here a few notes to supplement the slight accounts of the various instruments given throughout the book.

The sextant.

The sextant is pre-eminently a sea instrument, and it is not well adapted for use upon land, except in special circumstances.

The sextant will measure angles up to about 130° but not more. Held in the hand, it can be adjusted to allow for the motion of the ship in a way that becomes for the skilled observer almost automatic. It fulfils all the needs of the sailor, who measures elevations above the visible sea horizon, for determination of position, and angles between marks on land to clear hidden dangers. It gives results which are nominally correct to 10″ and actually correct to within half a minute of arc, which is just the degree of accuracy required in navigation. The principal systematic error to which the instrument is liable is the error of eccentricity of the graduated arc—the error which in the theodolite and other more precise instruments is eliminated by reading the circle at two opposite points, which is impossible in the sextant.

It is the possibility of this error developing undetected that makes it waste labour to equip a sextant with a stand, as has been done, and to try to make use of it upon land.

On land the sextant cannot give altitudes directly, because there is no visible sea horizon from which to measure. It is necessary to employ what is somewhat confusingly termed the "artificial horizon": a dish of mercury which forms a naturally level reflecting surface, or a reflector of blackened glass which can be set level by levelling screws and a bubble. The sextant then measures the angle between the direct and the reflected images of the sun or star, that is to say, double the altitude of the body. But angles greater than 130° cannot be measured. Hence when the sun at noon is higher than about 65°, which is very frequent in tropical and subtropical regions, the noon altitude of the sun cannot be taken on land with a sextant. This serious limitation is enough to condemn the use of the instrument on shore. Moreover, the sextant cannot measure azimuths except in a roundabout way, which involves a great deal of observation and unnecessary calculation.

If, then, it is necessary to retain the sextant as a subsidiary instrument for work on land, its use should be restricted to those cases in which it must be employed for reasons of secrecy. There are native tribes which would interfere with a surveyor working at an instrument mounted on a tripod who would not be so likely to notice a man lying down on his face taking sights with a sextant. It is said that native Indian surveyors have done good work in trans-frontier regions with a sextant disguised as a praying wheel.

The sextant has often been employed on polar expeditions, on account of its small size and lightness. The observations of such small altitudes as the sun has in spring in polar regions offers, however, peculiar difficulties when an artificial horizon must be employed; and it is now generally admitted that a small mountain theodolite should always be employed on such work. It has the great advantages that its errors are self-eliminating, and that it can be used with facility to measure azimuths. A modern mountain theodolite of the smallest size weighs no more than a sextant with artificial horizon, and the results that it gives are far more satisfactory.

Special forms of sextants have been developed for air navigation, with a bubble which is brought into the field of view by reflection, and serves to give an artificial horizon. Accelerations of the aeroplane displace the bubble, and the results are not accurate within several minutes of arc, but Commander Byrd's flight to the North Pole and back showed that in skilful hands the instrument can be made to work sufficiently well.

The theodolite.

For many years the ordinary surveyor's theodolite suffered from a want of intelligence in its design, of which traces still survive in the patterns manufactured especially for engineers. The most conspicuous of these is the four-screw levelling device, which inevitably strains the instrument, makes the screws work loose, is thoroughly bad in design, but is still very often made, and appeared in a recent catalogue of a leading English firm as the "Australian pattern".

The old-fashioned theodolite was effective only for horizontal angles, in which reversal is not much required. When it was found to be advantageous to make the instrument reversible, which requires that the telescope shall be able to pass through the frame, the new pattern was called on this account a "transit theodolite", which name still survives in the makers' catalogues, and in Canadian and American use: a stupid name which suggests that the theodolite can be used as a transit instrument.

The circles of the old-fashioned theodolite were read by verniers only, and verniers are awkward to read by day, almost impossible to illuminate properly for reading at night. A great improvement in theodolites was made in the application of the micrometer microscope, which increases the accuracy of reading the circle at least fivefold, and at the same time makes it much more simple and easy: a very unusual accompaniment of gain in precision. The five-inch micrometer theodolite established itself some thirty years ago as the standard instrument for survey of the second order, that is to say, in which geodetic accuracy is not required. The instrument is sufficiently portable for the employment upon boundary survey under the roughest conditions; and the facility with which it can be used for field astronomy makes it invaluable both in teaching and in actual use in the field.

The modern theodolite is especially adapted for the accurate measurement of vertical angles, in which the old instruments were very deficient. The improvements in construction which have led to this result are two: making the instrument reversible, already mentioned; and the mounting of a sensitive bubble upon the frame which carries the verniers or microscopes for reading the vertical circle. Since there is often some misconception as to the part played by these two devices, we may examine them shortly.

It must be remembered that there is a great difference between the measurement of horizontal and of vertical angles: the former are relative, the latter are absolute determinations. Or in other words, on the horizontal circle one measures the difference in bearing between one point and another; on the vertical circle one measures the elevation of a point above the horizon, which is not a visible object on which settings can be made, but must be defined by means of sensitive levels or bubbles mounted on the instrument. To measure an absolute elevation naturally demands that all the effects of errors of adjustment, collimation, zeros of microscopes, and so on, shall be eliminated; and this can be secured by reversing the instrument about a stable axis very nearly vertical. If the approximately vertical axis of the theodolite could be trusted to remain fixed during the series of settings and reversals of the instrument which take place in the course of a series of observations on an object, or even to remain fixed for the pair of observations, "face left and face right," which constitute a complete observation, then the reversal would eliminate all these errors of collimation and zero automatically, and nothing more would be required. But in practice the foundation of the instrument is not perfectly stable. Under the influence of the movements of the observer and the weight of the instrument itself, slight settling takes place between one setting and the next; and perhaps also the position of the microscopes on the frame changes a little by reason of change in the temperature. It is the function of the bubble mounted on the microscope arms to eliminate these changes, and the introduction of this device very much increased the accuracy of the results.

Personality in observation.

When the errors of the instrument are eliminated there remain the personal errors of the observer; and these also can be eliminated in great part by a proper use of the principle of reversal. An observer will have a tendency, quite unknown to him, and almost ineradicable, to make settings systematically high or low. This is called the personal error of bisection. It may be eliminated by combining observations in which the error comes in with opposite effects upon the quantity which is to be determined. Observations made north and south of the zenith will be affected with the same

$3\frac{1}{4}$-*inch Micrometer Theodolite (Watts) with mirror for reading bubble.*

Plate XX

1. *Geodetic and Universal Theodolites.*

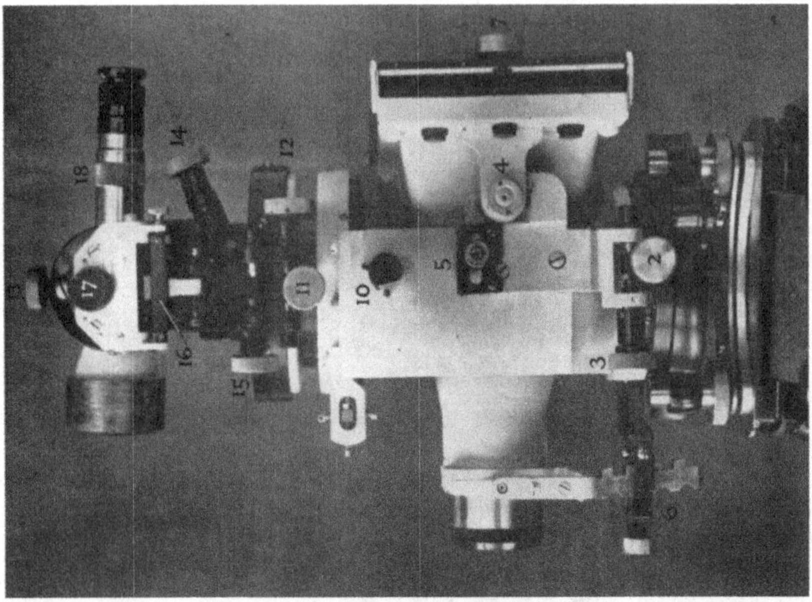

2. *Photo-Theodolite.*

error in altitude, but this will produce opposite errors in the resulting latitude. Similarly the personal errors in observations made east and west will have opposite effects upon determinations of time. Thus by preserving a balance between north and south or east and west stars the consequences of these errors will be eliminated in the mean.

A new type of theodolite.

The micrometer theodolite has the serious disadvantage that the observer must walk round the instrument to read the level and the microscopes, which takes time and is apt to disturb the tripod.

About the year 1921 a new type of theodolite was produced by Mr Heinrich Wild of Heerbrugg, Switzerland, with remarkable improvements:

1. The circles are etched on glass, the horizontal 95 mm. and the vertical only 50 mm. in diameter; yet fifteen of the former gave for the average angular error of a diameter—that is of the line joining an opposite pair of divisions—the very small quantity $0\overset{''}{.}398$.

2. Both circles are read in one telescope from the eyepiece of the instrument, by a parallel-plate micrometer which gives at one setting the mean of the two readings on opposite divisions. The drum of this micrometer is divided to single seconds and estimated to tenths.

3. Before reading the vertical circle the level is set central from the eye-end, so that there are no separate level readings to record and compute.

4. The same instrument, with one modification in the form of the vertical axis, is built as a photo-theodolite, which can be turned to rising or plunging views.

The design of this instrument marked the greatest advance that had ever been made at a single step, and those who first worked with it in the field were unanimous in saying that they would never revert to an instrument of the older type.

In the spring of 1927 Mr Wild produced the first model of a larger instrument of the same type, for geodetic work, with a micrometer in which the head is divided to tenths of seconds and estimated to hundredths. Whether such refinement can ever be used to advantage in the field remains to be proved: but to see for the first time a micrometer head divided to tenths of seconds is an unforgettable experience, and one must hesitate to say that the refinement is thrown away. It must at least reduce the necessity for many rounds of angles; and the small size and weight of these theodolites make them very economical of transport.

The photo-theodolite of this type belonging to the R.G.S. was employed by Major Kenneth Mason with great success in the Shaksgam Expedition of 1926 to the Karakoram Himalaya, where, packed in canvas travelling cases lined with sponge rubber sheet, it withstood very hard conditions. The instrument is fully described in the *Geog. Jour.* for April 1926.

A similar instrument embodying the principal patents of Mr Wild was produced about the same time by Messrs Zeiss, and successive small improvements have been made in both instruments in the light of field experience.

In the summer of 1926 the Chief of the Geographical Section of the General Staff invited the principal English makers to a conference at Tavistock, where a survey party were experimenting with the new instruments. As was anticipated, it proved no easy task to devise alternatives to the improvements covered by the Wild patents; but after some years' work the firm of Cooke, Troughton and Simms have produced the Tavistock Theodolite, described in the *Geographical Journal* for May 1931. For this instrument Instructor-Captain Baker devised an elegant method of obtaining at one setting of the micrometer the mean reading of the circle at two opposite points, which may well prove superior to the Wild device. But no definite conclusion on the respective merits of the two instruments is possible until the periodic error described by Major Cheetham in the above mentioned account has been traced to its source.

Only extended use in the field in every kind of bad condition of steaming heat, frost, or dust, can decide whether instruments with these complicated and rather inaccessible optical trains can be relied upon to function without breakdown. As was inevitable, the earlier models developed a few unexpected defects, and in particular showed themselves not very well planned for field astronomy. The electric illumination of the circles and micrometers has since been much improved, and it is now recognised that to get the best results artificial illumination of the circles and micrometers is necessary, even by day. The range of observation in zenith distance has been extended, star sights provided, and accessibility of the internal parts improved. There can be no question that the merits of the new form are thoroughly established.

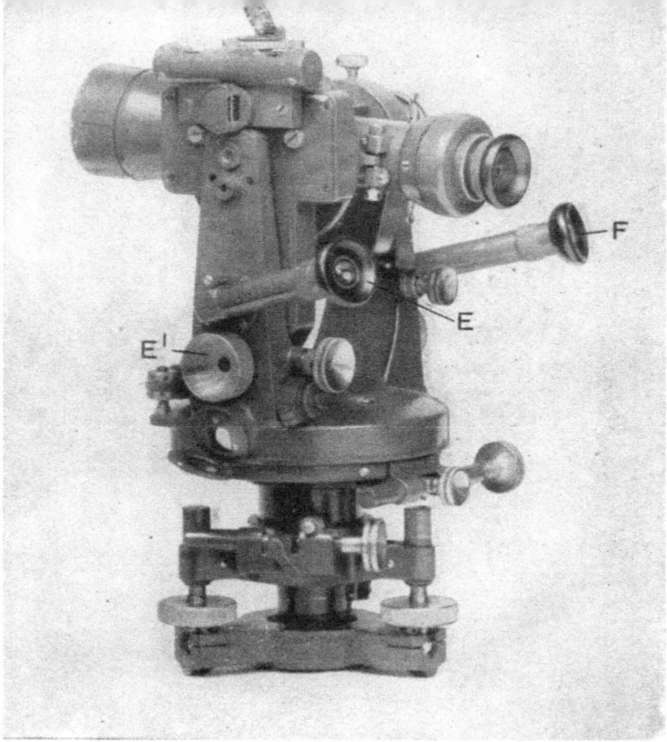

(From paper by Major Cheetham, R.E. *Geogr. Jour.* LXXVII, 442, May 1931)

The Tavistock Theodolite (*Cooke, Troughton and Simms*).

E, E′ Eyepiece and Micrometer Vertical Circle. F, F′ Eyepiece and Micrometer Horizontal Circle.

Plate XXII

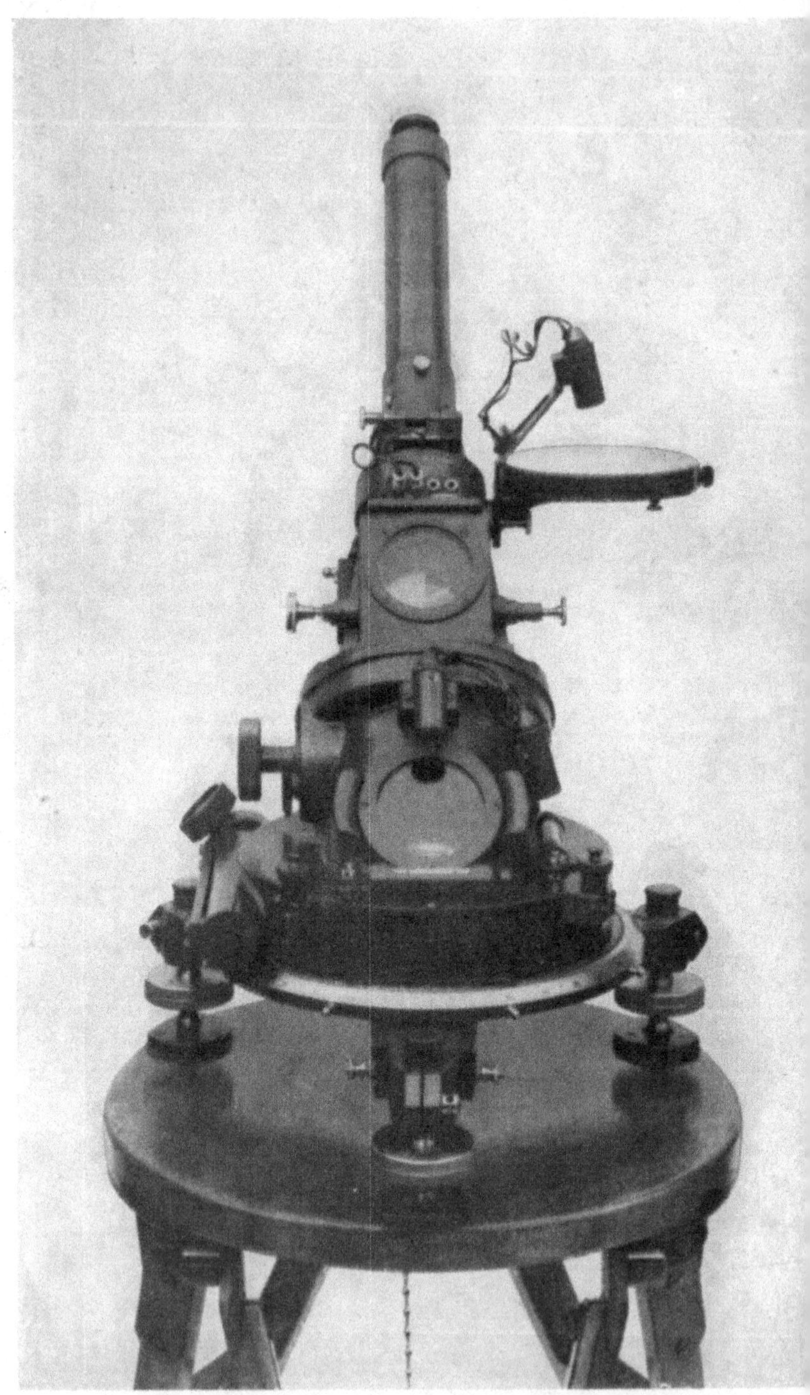

(*Described in paper by Instructor-Captain T. Y. Baker, R.N. Geogr. Jour.* LXXVII, 429, *May* 1

The 45° *Prismatic Astrolabe with pentagonal prism and illuminated compass for s*

The prismatic astrolabe.

The "astrolabe à prisme" was invented by MM. Claude and Driencourt some thirty years ago, and was used by the French for some years before its merits were generally recognised. Its essential parts are a horizontal telescope, a sixty-degree prism fixed immediately outside the objective with the refracting edges horizontal and the face next the objective vertical, and a dish of mercury beyond and a little lower than the prism. When a star is at an elevation of 60° its light passes normally through the upper inclined face, is totally reflected at the lower inclined face, and enters the lower half of the objective horizontally. Light from the star is also reflected up from the mercury through the lower face and totally reflected by the upper face into the upper half of the objective. The two beams then combine to form a single image of the star. And it is clear that just before this, when the star has not quite reached the elevation 60° two images of the star will be seen approaching one another in the field of view. The instant of contact can be recorded by stop-watch or on the chronograph.

The instrument is relatively insensible to maladjustment: that is to say, a small error in the horizontality of the telescope or of the refracting edge of the prism, has very little effect on the time of coincidence of the images.

It is shown in the *Handbook of the Prismatic Astrolabe* by Ball and Knox-Shaw that stars should be selected which arrive at the elevation 60° in azimuths as near as may be to 45°, 135°, 225° and 315°, and tables are provided for choosing these stars. The reduction, by the method of position lines, is semi-graphic, which has the great advantage that an erroneous observation can be detected immediately. There are no divided circles, except a rough setting circle in azimuth, and if the refracting angle of the prism is not precisely 60° the error is automatically eliminated.

The instrument has an optical disadvantage in that the images are produced by one half only of the objective for each, and are consequently slightly elongated and chromatic: but this does not seem to give trouble in practice, and there seems to be no doubt that the more a man works with the instrument the better he likes it. It will probably displace the theodolite for observation of time and latitude, except when only one instrument can be carried. In that case it must be the theodolite, since the prismatic astrolabe cannot determine azimuth. But Mr E. A. Reeves has introduced a prismatic attachment to the theodolite which can convert it at will into a prismatic astrolabe, and this device was used with great success by Mr Weld Arnold on Dr Hamilton Rice's expedition to the Rio Branco.

A more fundamental optical defect in this pattern is that the prism divides the objective at right angles to the direction of measurement, and that consequently a slight change in the focal setting of the eyepiece alters the time of apparent coincidence. Instructor-Captain Baker has therefore recently designed a new form of the astrolabe, in which the telescope looks down at 45° through a pentagonal prism into the mercury, and stars are observed at altitude 45° instead of 60°, which gives more stars available. By additional thin prisms one image of the star is divided into two, and the other image passes between them, which gives better observation; and three observations of each star can be made instead of one. The instrument is fully described in the *Geographical Journal* for May 1931 and the methods of reduction discussed. With this instrument a good observer can obtain his position within 1″ from twelve stars in a couple of hours' observation.

Levels.

Very great improvements have been made in the construction of precise levelling instruments during recent years. In the older instruments it was necessary first to adjust the foot screws so that the bubble was in the centre of its run, and then to walk round to the eyepiece and make the reading on the staff. In the meanwhile the bubble had probably shifted. In the modern type of level invented by Mr Heinrich Wild and introduced by the firm of Zeiss it is possible to see both ends of the bubble by a second telescope fitted alongside the principal, and read with the other eye. The final adjustment of the bubble is thus made immediately before reading the staff, and without changing position. Further, all the essential parts of the instrument may be constructed of invar, so that the effects of temperature are almost eliminated.

Three parallel horizontal wires are fitted, and the staff is read against all three. This helps to eliminate accidental errors of reading, and also provides, on the principle of the tacheometer, for

Plate XXIII

ve-inch Casella Theodolite fitted with Reeves' Prismatic Astrolabe Attachment.

Plate XXIV

1. *Straining trestle.*

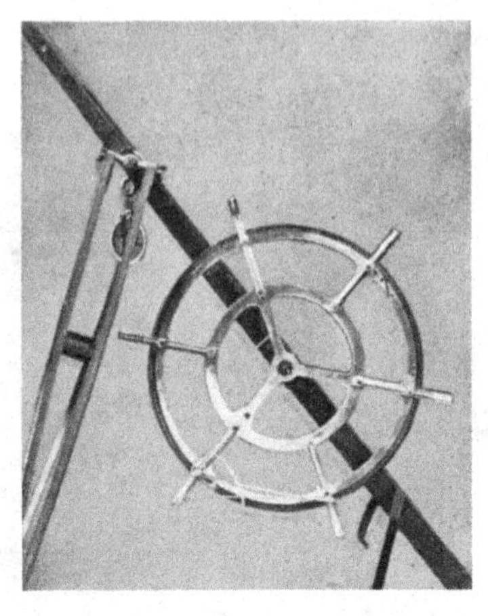

2. *Tape on drum.*

3. *Mark on tripod.*

4. *Alignment sight on tripod.*

Invar-tape Base Apparatus made in the Cambridge Observatory workshop about 1911 by Mr Gordon Dobson.

the control of the equality of the distances between the forward and the back staves.

The staves are graduated on both faces, the second graduation being from an arbitrary zero, so that the second set of readings is not a close repetition of the figures of the first. And special precautions are taken to control at frequent intervals the lengths and the graduation of the staves.

A further great improvement is the parallel plate micrometer mounted before the object-glass, by which it is possible to measure directly, and independent of the distance, the difference of height between the axis of collimation and the division of the staff. The wires can thus be set upon a division: it is no longer necessary to estimate tenths of a division: and the staff can be made a fine uniformly graduated scale, instead of the confusing old pattern.

Invar wire or tape base apparatus.

As the result of experience in the field many improvements have been made on the original design of M. Guillaume, of the International Bureau of Weights and Measures at Breteuil. An apparatus designed and made by Dr Gordon Dobson, F.R.S., when he was a student in the Cambridge School of Geography about 1911, serves very well to illustrate the essentials.

The straining trestle has, as is now usual, one leg prolonged to rest on the shoulder of the operator, so that he can take the weight of the wire while he adjusts the other two legs. The novel feature of the design is the swivelling hinge of these legs, so that the tripod is capable of some lateral motion without taking the points of the legs from the ground. This is very convenient in adjusting the wire to the line of the tripod marks.

For convenience and rapidity of work it is essential that the tripod mark should be adjustable over a range of a foot without moving the tripod itself; for on rough ground it is not possible to set the tripod up level at any exact point desired, nor to move it readily by small amounts. The head of the Cambridge form of tripod is a skeleton triangle; and the post for the mark is carried through two battens, one above and the other below the triangle. A wing nut at the base of the post tightens the two battens and holds them, or allows a range of motion and possibility of clamping at any point over a circle of about a foot in diameter.

Geodetic pendulums.

The form of apparatus in general use is due to Major-General von Sterneck. It consists of a set of several small pendulums adjusted to swing in very slightly more than half a second. The pendulums are made of brass, heavily gilded to avoid change by corrosion. They rest by delicate agate knife-edges on an agate plane carried by the stand. Each pendulum carries a small vertical mirror on its head, just above the line of the knife-edges. There is an arrangement for giving the pendulums a small displacement and then allowing them to swing freely; the arc of vibration is less than half a degree. They are set swinging and gradually come to rest.

A dummy pendulum similar to the others contains a thermometer, to determine the temperature of the pendulums. The stand is massive; but it is not possible to consider it as perfectly free from flexure and vibration. There is therefore a special arrangement attached to the stand to enable the correction for flexure to be determined. The principle of this determination is elegant. If two pendulums hang side by side and one is set swinging, while the other is initially at rest, the latter will gradually take up the oscillations of the former, unless the stand is perfectly rigid. The flexure of the stand is thus determined from the rate at which the first pendulum influences the second.

The periods of vibration of the pendulums are found by comparing them with the beats of a half-seconds clock or chronometer. Each half-second the clock momentarily opens the shutter before the slit of the "flash-box", and the illuminated slit is reflected in the vertical mirror on the pendulum, and viewed in a small telescope on top of the flash-box.

Thus a line of light is seen in the field of the telescope every half-second; and, as the clock gradually gains on the pendulum, the position of the line moves down in the field, and the time when it coincides with a central horizontal wire can be determined. The interval between one coincidence and the next is the interval in which the pendulum loses one swing on the clock. The observations being continued over a period of several hours, the comparison between the clock and the pendulum becomes of great precision, with a probable error of only three or four parts in ten million.

This apparatus is fairly portable, and the necessary observations for the determination of the value of gravity at a single station may be made in a couple of periods of three or four hours, generally chosen so that one is at night and one by day.

It used to take several days to get the clock rated by star-transits, even with clear weather, and there was often great delay from cloud. Now that the time can be obtained by wireless signals very much more accurately than it could be observed in the field, and quite irrespective of the weather, gravity survey is much simplified.

A very great improvement in the theory and design of geödetic pendulums has been made by Dr Vening Meinesz, originally to eliminate the vibrations in the insecure soil of Holland, but soon adapted to use in a submerged submarine. If two pendulums are swung in the same plane in opposite phase, the effect of a displacement of the stand is equal and opposite on the two, or is eliminated from their mean by a simple geometrical discussion. Dr Meinesz used four pendulums swinging in pairs in planes at right angles, and all recorded photographically by beams of light reflected from small mirrors on the pendulum heads. When the rolling of the submarine did not exceed one degree, the observations had an accuracy about half that obtainable on land. Dr Meinesz finds that the most serious cause of uncertainty remaining is the difficulty of determining the east and west component of the current, which changes the centrifugal force of the Earth's rotation. A velocity of one knot corresponds to a change of ·000 007 5 of the value of gravity. (See *Geog. Jour.* LXV, 501, June 1925.)

The principle first adopted in the above apparatus has been applied by Sir Gerald Lenox-Conyngham in the new pendulum apparatus built by the Cambridge Instrument Company for the Geodetic Institute at Cambridge. The beam of light is observed after successive reflection at two mirrors carried on a pair of pendulums swinging in opposite phases in the same plane, the observations being made by the flash-box method as in the old apparatus. All difficulty about instability of the stand is thus eliminated. Still more recent practice does away with the clock and flash-box at the station, and obtains the flashes from the oscillations of a standard pendulum at the base station, transmitting signals by radio to the field station.

Surveying cameras.

A recent official textbook speaks of an "ordinary camera fitted with levels" as being all that is required for photographic survey on the "Canadian Method". This is dangerous doctrine. The following points are essential:

1. The camera must be rigid and work at fixed focus, and the focal length of the lens must be known accurately.
2. The plate carrier must be designed so that the plate is brought up hard against three stops accurately worked to a plane which is perpendicular to the optical axis, and vertical when the camera is set up by its levels.
3. The stops must be accurately worked to give the horizontal and vertical lines on the plate through the point where the optical axis cuts the plate. This is best secured by piercing the stops with small holes that photograph as round black dots on the plate; and there should be two in each stop, to avoid as far as possible the loss of a fiducial mark by its coincidence with a dark marking in the landscape.
4. The camera must be mounted on a steady tripod with levelling screws like those of a theodolite, and the whole instrument most carefully built and adjusted so that the above conditions shall remain fulfilled during rough use in the field.

The surveying camera will usually be set up at a point which is to be determined with the theodolite by triangulation or resection; and there are clear advantages in combining camera and theodolite in one instrument, the photo-theodolite, to save transport, and time in setting up.

The Stereo-comparator of Pulfrich.

If two photographs taken on parallel axes from the ends of a measured base are combined in a stereoscope, the ground is seen in relief. The stereo-comparator is a special form of stereoscope designed to measure the relative displacement of corresponding points in the two pictures, and to deduce from this displacement the distance of the point on the ground from the base. This being known, it is easy to deduce the height of the point above one end of the base from the measured angle of elevation. Each telescope of the instrument contains in the focal plane a small black pointer, and the first adjustment is to move the eyepieces laterally over the marks until their distance suits the interocular distance of the observer; the two pointers then appear as one. Next adjust the plates until the fiducial marks combine with the pointers. If now we traverse both plates together, the pointer appears to lie behind the landscape, which is seen in strong relief: but by moving one plate relatively to the other, the pointer can apparently be brought forward into contact with any desired feature of the ground, and the amount of relative motion required to effect this contact is a measure of the distance of the feature. The effect is difficult to explain, for the pointer seems to advance and retire, though in reality the pointer is fixed, and the motion is in one picture relative to the other. The common explanation of stereoscopic vision, that it is connected with an appreciation of the degree of convergence of the axes of the eyes, is evidently not sufficient here, for the convergence is defined by the pointers under the two eyepieces, and remains constant. But if the eyes of the observer are fairly well matched the effect is very striking, and the measure is very precise.

With the simple stereo-comparator above described the distance and height of any feature can be measured, and the results plotted by hand on the drawing. The great advantage over the method of photographic survey by measuring plates separately is that no time is lost in identifying corresponding points on the photographs. The identification is automatic, and the measure may be made on any desired part of the field, whether it is marked by a conspicuous feature or not, and this is an enormous advantage.

The stereo-comparator was brought out by Pulfrich about 1904, and a good deal of stereographic survey has been done with it: e.g. the survey of Mount Olympus by Marcel Kurz. It is accurate, but slow, since the horizontal co-ordinates and the height of each point must be calculated from the measures and plotted by hand.

Stereo-plotters.

The method has, however, much greater possibilities than this. By an elaboration of the stereo-comparator it can be made to plot the map itself, and even to trace the contours.

The first machine to realise a part of these possibilities was constructed by Captain F. Vivian Thompson, R.E., of the School of Military Engineering, Chatham, and described by him in the *Geog. Jour.* (May 1908). It gave automatically the horizontal position of the point, and its height above one end of the base, but it did not actually draw the map, and it could not trace the contours. The complete realisation of the method was made about

1911 by Lieutenant von Orel of the Military Geographical Institute of Vienna, in co-operation with Messrs Zeiss of Jena. They have made several models with successive improvements. The model of 1911 is described by M. Paul Corbin of Paris in *Revue générale des Sciences pures et appliquées*, 30 March 1914; the model of 1914 installed in Paris by M. Corbin at the Société française de Stéréotopographie is described in *Geog. Jour.* April 1922: and a still later model of 1920 is known only by its patent specification, consisting mainly of an elaborate trigonometrical formula which may embrace a great deal. This model was evidently intended to deal with plates taken in any plane, especially air photographs, but it does not seem to have come into use, and the several types of machine since developed for dealing with oblique photographs have been on a different principle, first employed by the late Professor Kapteyn in his parallactic micrometer for measuring star photographs: the principle of the Bildmesstheodolit or Photogoniometer. The plate is placed in a camera identical optically with the camera which took it in the field. Rays drawn back from the plate through the first principal point of the objective have the same geometrical relations one to another as had the rays that took the photographs originally. Hence if we could set up a theodolite to observe the plate through the camera objective we could survey the photograph precisely as we should survey the country by a camera set up at the theodolite station. The geometrical conditions are strictly similar, and the observed angles would bear the same relations to one another.

If it were a question of observing angles from one photograph we could keep the camera at rest and move the observing telescope. But since we want to study a pair of photographs with a binocular telescope we must keep the telescope fixed, and turn the cameras about instead. There are some six machines embodying different solutions of the problem, how to turn these cameras about so that their axes reproduce the geometry of the rays making the images. If that can be done it is evident in a general way that by gearing a pencil carriage to the mechanism that turns the cameras about, one can plot the map, including contours, from pairs of photographs; the difficulty is to give the cameras the great range of motion required without fouling.

The Autocartograph of Hugershoff and the Stereoplanigraph of

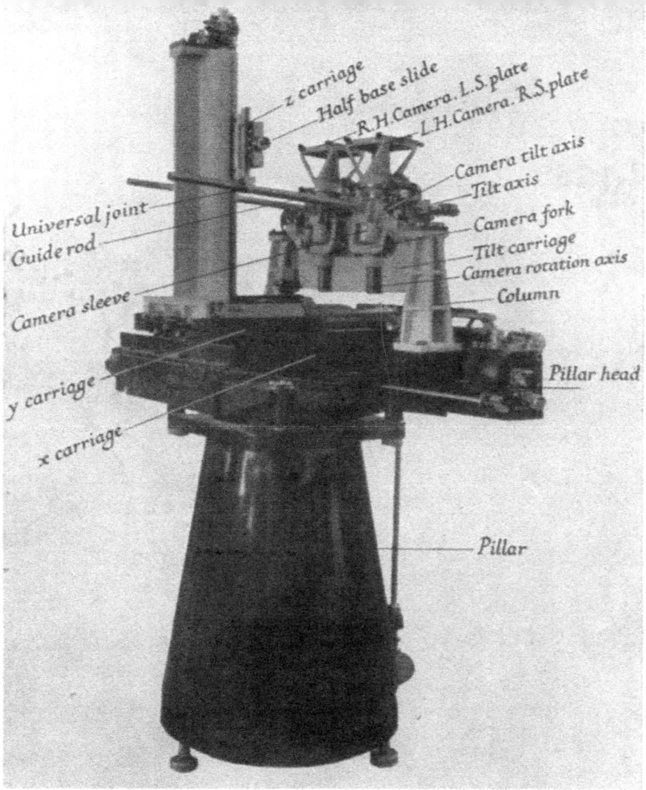

The Wild Autograph A 1 (*drawing table removed*).

Plate XXVI

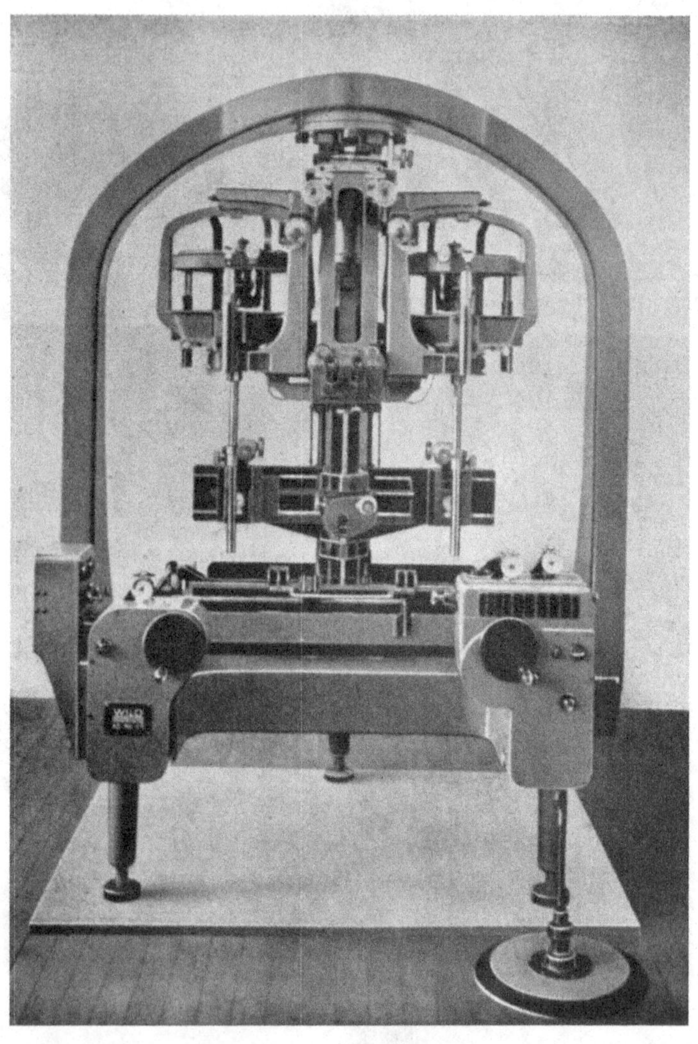

The Wild Autograph model A 5. Front view of the plotting machine.

Zeiss are designed especially for plotting pairs of air photographs. The first is described in *Geog. Jour.* April 1922 by Lieut.-Colonel MacLeod. The Autograph of Wild is designed primarily for plates taken at ground stations, including those with a large range of tilt or of obliquity to the base, or for air photographs whose axes are within 40° of horizontal; though by a simple interchange of gears the Autograph is equally well adapted for air survey with plates taken within 40° of the vertical. The machine is in regular use by the Swiss Federal Surveys, and by several commercial firms. It is the most compact and the least expensive of the several machines, and a fairly full description will be found in the *Geog. Jour.* for October 1927.

The Barr and Stroud Photogrammetric Plotter is described by Captain Hotine in the *Geographical Journal* for February 1930. For fuller information the reader may refer to the recent works by von Gruber and Hotine.

The Autograph.

Consider the simplest case, of plates taken normal to the base, and with axes horizontal. The plates are adjusted in skeleton cameras identical with the field camera, and viewed by a binocular telescope pointing into 45 degree reflecting prisms immediately below the camera objectives. The cameras can be turned with their prisms round vertical axes, or can be tilted about horizontal axes. They are controlled by guide rods moving in sleeves carried on the base slides of the z-carriage, which is moved up and down by the foot-wheel. The vertical z-column is moved on the y-carriage towards or away from the observer, and the y-carriage is moved right or left on the x-carriage, both movements by hand-wheels.

When the camera is twisted the line of sight of the observing telescope sweeps along the plate parallel to the x co-ordinate, and when it is tilted, parallel to the y co-ordinate. The sleeves of the guide rods are attached to the base slides, which are set outwards from zero each to half the length of the base on the desired scale of the map, and the right-hand base slide can also be adjusted vertically to allow for difference of height at the ends of the base.

Suppose a point P in the field, bearing α to the right of the normal to the base, and with angular elevation β. The co-ordinates of its image upon the plate taken normal to the base are

$$p = f \tan \alpha, \qquad q = f \sec \alpha \tan \beta.$$

Let the hand and foot wheels be turned until the guide rod takes up the same angles relative to the instrument base as the above ray does to the field base. Then the camera attached to the guide rod is twisted through α and tilted through β, and it is easy to show that the telescope points to a point on the plate whose co-ordinates are

$$p' = f \tan \alpha \sec \beta, \qquad q' = f \tan \beta.$$

This "false point" and the "true point" are equidistant from the centre of the plate, and therefore the telescope can be made to set on the true point by rotating the camera through an angle ρ where

$$\tan \rho = \frac{\sin \alpha \sin \beta}{\cos \alpha + \cos \beta}$$

and this rotation is effected by a system of cams which are partly described in the paper quoted above, but of which the details remain the secret of the inventor. The compactness and mechanical success of the Autograph are due to the invention of this cam system.

When, therefore, the controls are operated so that the floating marks are brought into stereoscopic fusion with a point in the picture, the guide rods reproduce the geometry of the rays which took the plates, and a pencil moving on a carriage geared to the controls plots the point under observation. Moreover, if the z-control is set to a given height, and the mark made to travel over the picture in apparent contact with the ground, the pencil traces the contour at that height.

So much for the simple case. The reader must be referred to the fuller description for the account of how the machine is made to deal with plates oblique to the base, or with inclined or convergent axes.

So far as can be seen at present, it seems likely that some form of photogoniometer is necessary for the solution of the general case, with plates inclined to the vertical. But such machines must always be complex and expensive, and the simpler geometry of the plate with horizontal axis is worth keeping in mind, because a simple machine which could be used in the field is very desirable. We therefore retain the account of the Stereo-autograph of von Orel and Zeiss which was given in the second edition of this book.

Geometrical theory of the Stereo-autograph.

Though the stereo-autograph as made by Zeiss is very complicated, and the action of its inter-connected levers difficult to follow, the geometry underlying them is really very simple, and may some day be realised in a simpler way.

In the figure S_1 and S_2 are projections on the horizontal of the two stations at which the photographs are taken (S_1S_2 being called the base) and M is the point to be fixed, which is projected on the two photographs at M_1 and M_2. The photographs are in parallel vertical planes inclined at an angle θ to the base.

Transfer the optical centre S_2 to S_3, so that the second plate lies in the plane of the first plate, the distance S_1S_3 being arbitrary.

Draw S_3R parallel to M_1S_1M; MR parallel to S_1S_3; M_3S_3Q parallel to M_2S_2M; and RQ parallel to S_1S_2. Join MQ.

The triangles MS_1S_2, S_3RQ are equal, since their sides are parallel and $S_1M = RS_3$. Hence $QR = S_1S_2$ and $QRM = \theta$, the inclination of plates to base.

Hence the triangle QRM has its side $QR =$ the base; angle $QRM = \theta$; and side RM equal and parallel to S_1S_3: that is, its size and shape and orientation are independent of the position of M.

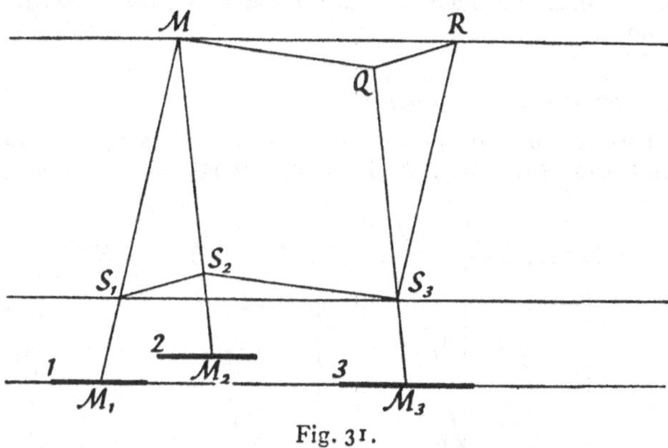

Fig. 31.

If then in the above figure we had given only the points S_1, S_3 and M_1, M_3, we could find the position of M by applying the invariable triangle to M_1S_1 and M_3S_3 produced, so that M and Q lie respectively on these rays, and MR is parallel to S_1S_3.

This is the simple geometrical principle underlying the action of the stereo-autograph.

Mechanical construction.

The optical parts are simply the well-known stereo-comparator of Pulfrich. The two plates (numbered 1 and 3 to correspond with the geometrical figure above) are on a carriage moved horizontally by the hand-wheel labelled Direction, along the line S_1S_3. Plate 3 has an independent movement relative to Plate 1, along the same line, on an upper carriage.

The plates are viewed by two compound microscopes whose objectives O_1 and O_2 are at fixed distance apart S_1S_3, and whose eyepieces E_1, E_2 receive the light from O_1, O_2 through three reflecting prisms in each microscope.

The odd number of reversals by reflexion, combined with the normal inversion of the field in the microscope, give an image of each plate right way up but reversed right and left: if then the negatives are inserted film upwards (i.e. reversed R. and L.) the resulting view is correct, except that it is a negative, black for white.

Under the eyepieces are the pointers, and the plates are adjusted in their carriers so that the images of the fiducial marks fall simultaneously upon them. Plate 3 has an additional independent movement at right angles to S_1S_3.

The whole optical system can be moved parallel to the plane of the plates along the line GH perpendicular to S_1S_3. Thus by moving the plates under the microscopes in the horizontal sense, and by moving the microscopes over the plates in the vertical sense, the whole plates can be examined.

The plotting of horizontal detail.

Two bars A and B are pivoted about fixed centres s_1, s_3 so that s_1s_3 is equal and parallel to S_1S_3. A sleeve m_1 through which bar A can slide,

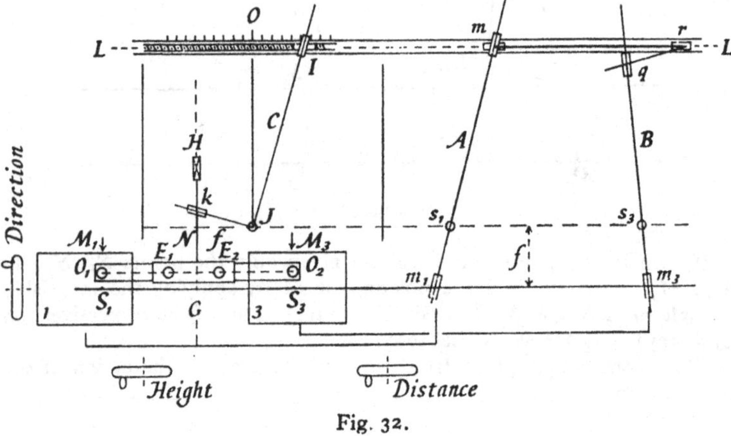

Fig. 32.

is linked to the carriage of the two plates, so that m_1 moves with plate 1 along the line S_1S_3. Similarly the sleeve m_3 on bar B is linked to the upper plate carriage on which plate 3 moves relative to plate 1. The line of motion $S_1S_3m_1m_3$ is parallel to s_1s_3 and distant from it f, the focal length of the objective with which the plates are taken.

A bar LL' can be moved by the hand-wheel marked Distance always parallel to itself, in the plane of the figure. A bar mr of fixed length equal S_1S_3 or s_1s_3 can slide along LL'; it is connected with the bar A by a pivoted double sleeve.

A bar rq is clamped to mr at r, at an adjustable angle $= \theta$, the inclination of the plates to the base on which they were taken. And the length rq is adjustable, so that rq is the length of the base on the scale of the desired map. A sleeve pivoted to rq at q slides on B.

If now the direction handle is turned to move the plates together under the microscopes, m_1 is displaced horizontally in the same direction and by the same amount. The bar A turns about s_1 and pushes mr in the opposite direction along LL'. This carries q with it and turns B about s_3, displacing m_3 by an amount not quite the same as m_1 (unless θ is zero) and consequently displacing plate 3, which is linked to m_3, a little with the upper carriage, relative to plate 1. Note that if the two plates were on independent carriages (as they are in the 1911 model) the result is just the same, except that the whole movement of plate 3, instead of only its differential movement, is transmitted through the train of levers, and this is not quite so good mechanically. Note also that, by the similarity of Figs. 31 and 32, the differential movement of 3 is exactly right to maintain the setting of the R.H. microscope on the same point as the L.H., if the point moves along the line MR at a fixed distance from S_1S_3. MS_2M_2 is parallel to QS_3M_3, and similarly divided. Now displace LL' parallel to itself by the distance hand-wheel. The plate 1 and the bar A remain fixed; mrq slides along LL', and transmits through B and m_3 a motion to plate 3, relative to 1, with the result that the apparently single pointer seems to advance or recede in the field of view. When the pointer is thus set upon an object in the field, s_1m is its direction and distance on the map from s_1, the station at which plate 1 was taken, and a pencil underneath m will plot on the map the position of M. Thus the horizontal detail is drawn. We may either plot individual points, as corners of houses, or we may make the pointer traverse a road, and draw it continuously on the map.

Plotting contours.

So much for the horizontal detail. We have now to consider the question of heights.

The centre of the microscope system moves in the vertical sense along the line GH. A rod attached to the carriage slides in a fixed bearing H and carries a smooth sleeve pivoted to it at k. A rod Jk turns about a pillar fixed at J in the line of s_1s_3, distant f from GH, and slides in the sleeve k.

A rod IJ fixed at right angles to Jk passes through a sleeve at I jointed to a nut which is traversed along the bar LL' by a long screw turned by bevel-gear (not shown) from the hand-wheel marked Height, or from a foot-wheel connected with it. A divided scale with zero O adjusted so that JO is parallel to GH, measures the displacement of I right or left, corresponding to the displacement kN of the microscopes above or below the optical centres of the plates: that is, to the height co-ordinate of the point on the plate. But the triangle JOI is similar to JNk; and JO is the distance co-ordinate of the point on the scale of the map. Hence OI is the height co-ordinate on the same scale; and if the divided scale is the same as the scale on which the base rq is set, the readings OI will give directly the height of the point M above S_1.

The operation of this system of levers requires a little study. If the height wheel is turned OI is varied, the system IJk rotates about J, and the microscopes are moved in the vertical sense over the plates. This provides for the measurement of all heights in a co-ordinate plane parallel to the plane of the plates as taken, and distant by an amount corresponding to the distance between LL' and s_1s_3.

If on the other hand the height wheel is left untouched, OI remains unaltered in length. But if the distance wheel is turned, and LL' is moved away from Js_1s_3 the rod C slides in the sleeve at I; the angle IJO is decreased, and the microscopes travel downwards over the plates, by an amount corresponding precisely to the diminution of vertical co-ordinate on the plate as a point of given height above the base is removed further from it.

By this elegant construction, therefore, the height of any point in the field above the base point S_1 may be measured on the scale OI. But alternatively, if OI is set the vertical interval of any desired contour above or below S_1, and the height wheel is left untouched, the pointer can, by moving the direction and distance wheels, be brought into contact successively with all points on the contour, and m describes the contour. Moreover, if a print from plate 1 is adjusted on a table to the left of the plate, a tracing arm carried on the microscope will trace on the print the contours in perspective: a remarkably simple and sometimes valuable property.

We have spoken of the map as traced by a point below m. This requires a drawing inaccessibly placed under the machine. In the later models a stiff bar is rigidly attached below LL' and extended to the left, to carry a pencil over a drawing board quite clear of the machine.

One more beautiful property must be mentioned. If the plates 1 and 2 are not parallel when they are taken at S_1 and S_2, but their axes converge somewhat, they will still exhibit a relief when examined in the stereo-comparator, though the perspective will be false. To make a map from such a pair of plates it is necessary only to cut the arm B in such a way that its two parts qs_3 and s_3m_3 can be clamped at an angle equal to the angle between the planes of the two plates when they were taken. But the proof of this remarkable though very simple theorem must be left to the reader.

Chapter XIV

ADDITIONS AND CORRECTIONS TO THE PRECEDING CHAPTERS

Numbers prefixed refer back to the pages amended.

6. *The Este World Map* was published in four sheets in colour in 1934. If these are mounted on a circular panel of five-ply wood with a border of gold paint they make a very decorative object.

9. *The Molyneux Globes* have happily survived the damage that befell the Middle Temple Library by enemy action, and with the use of polaroid screens it may one day be possible to reproduce them.

13. *Conventional signs* must be adapted to the country, but may if they are easy to draw and clear to read be used for different features in different countries: see Notes by Lt.-Col. F. J. Salmon, *Geog. Jour.* LXXXIX, 50.

For Physiographical Maps and Cartograms whole series of new signs have been developed, for which see *General Cartography* by Erwin Raisz, and the review in *Geog. Jour.* XCVI, 351–5.

17. On the importance of *Lock and Sluice signs* in fen country see the discussion on O.S. Maps in *Geog. Jour.* LXXXVII, 321–2.

20. The Fifth (RELIEF) Edition of the O.S. one-inch map has been discontinued, to the regret of some and the satisfaction of others. For an appreciation see *Geog. Jour.* LXXVIII, 353, and for depreciation LXXIX, 79.

21. *Oblique Hill-shading* from a north-west light probably derives from the convenience of a light from the left front in working. The system is now abandoned also by the Ordnance Survey, but may be studied on recent Norwegian maps, and on the Karakoram map of the R.G.S.

22. *Contours* should be at uniform intervals on large-scale maps, but this is not always possible in mountainous country on smaller scales. On a contoured map of Asia the 50 and 100 metre contours are essential, especially in the coastal regions, even though the others may be 200, 500, 1000, and so up by thousands.

24. Contours of the beauty shown on Plate VI, p. 21, are not unnecessary luxury; for the proper study of land forms they are essential, and only stereographic survey can give them. They are even more important in the study of moving objects: accretion of talus slopes, advance or retreat of glaciers, or the forms of waves and of dunes.

26. *Layer-colouring* should proceed by very small steps, difficult to achieve by lithography and tints by transferred rulings. It is hoped that photogravure printing may be made to give many more gradations of a single colour.

27. *Mountain drawing* has been much developed of late in the United States for the physiographic map: see note 13 above for references. For *Combined systems of showing relief* study the now abandoned Fifth Relief Edition of the O.S. one-inch, the Norwegian tourist maps, and the Karakoram map of the R.G.S.

35. *The National Grid* of 5000 yard squares is to be superseded by the Metric Grid, for which a small table is given in *Geog. Jour.* XCVIII, 117. For the grid in yards see papers by Brigadier H. St J. L. Winterbotham in *Geog. Jour.* LXIII, 491, and LXXXII, 42; for arguments on the respective merits of yard and metre see the discussion in LXXXVII, 308–21, and the note on pp. 324–7, which deals also with numbering the grid. For discussion of the decision to adopt the metric grid see *Geog. Jour.* XCIII, 314–32.

A single grid cannot conveniently cover more than a few degrees of longitude, or the inclination of the meridians and the grid lines becomes excessive. One can have a single grid for Great Britain and include Ireland, but could not extend the map far eastward without breaking the grid and starting with a new central meridian and origin. This involves the difficulty of overlapping grids on some sheets unless the sheets have been laid out originally to suit the grid system and have the change of grid at a bounding meridian of the map. This difficulty is well illustrated by the German 1/100,000 and 1/300,000 which were laid out on the polyhedric projection before grids were thought of. A Transverse Mercator grid has been adjusted to them, but the junctions of successive sections of the grid are very awkward.

47. *Quill written names* have not yet won their way into official maps but are now used almost entirely on maps published by the

Royal Geographical Society. An extensive map of Europe and the Middle East on the scale of 1/11,000,000, made for the British Council, will be the first to use written names on an elaborate map printed in six colours. To see how legible minute quill figures in old style may be, study the map of Southern Arabia which illustrates a paper by Mr H. St J. Philby in the *Geog. Jour.* for August 1938.

50. *The frame* may sometimes be omitted with advantage, especially when meridians and parallels are strongly inclined and difficult to number. One can make good use of a clear inch outside the engraved surface of the map to number graticules and grids; to show direction to places just off the map and rivers which run off the map and on again; figures for variable scale along parallels, and other technical information.

To discard the frame is especially important when the projection is not symmetrical upon the sheet, as in maps recently drawn by the R.G.S.

53. The P.C.G.N. has recently made important decisions tending to discourage conventional names. In all future lists the names used by the local authority will constitute the principal entry; if there is a well-established conventional name, that name will also appear in the list with cross-reference to the principal entry. Another important decision is that systems of transliteration into the English form of the Roman alphabet, devised by states using other alphabets, shall not prevail over better systems already in British use, but that forms of the Roman alphabet officially adopted by foreign states formerly using other alphabets must be recognised. This decision allowed the Committee to adhere to the Hepburn system of romanising Japanese names, although the Japanese Government had issued a decree in 1937 for the gradual introduction of the Nipponsiki system. It also compelled the Committee to adopt the Turkish official romanisation, which has among other peculiarities two letters *I* involving a lower case *ı* without a dot and a capital *İ* with a dot. See "Recent Decisions of the P.C.G.N.", *Geog. Jour.* XCI, 158.

55. The Karakoram Map published by the R.G.S. in 1940 embodied the decisions of the Karakoram Conference on range

and peak names, about which there had been much controversy. The Conference Report was published in the *Geog. Jour.* XCI, 129–52, prefaced by an explanatory memorandum by Professor Kenneth Mason. The Report and note throw much light on the difficulties of systematic nomenclature in a region with many languages.

Rivalry in names may arise between two states active in exploring the same area such as the Antarctic. Until recently there were two sets of names, largely different, for a wide stretch of Antarctic coastline in the Australian sector. In the *Geog. Jour.* XCIV, following p. 272, is a map to illustrate a paper by Mr Lars Christensen, which shows in separate colours the overlapping claims to discovery and the different names used in Norwegian maps, brought together for the first time on a single map as material for consideration of the questions involved. The Government of the Commonwealth of Australia, the dominant authority for names in the Australian sector, has published a good map of the whole Antarctic, giving friendly and generous consideration to the claims of the Norwegian explorers.

58. *Drawing on card* is now being superseded by drawing on paper mounted on thin aluminium sheet or on zinc and aluminium plates enamelled with a dull finish. Drawing on stone is now little used for maps in the *Geog. Jour.* The finished drawings on mounted paper or Bristol board are reproduced by photo-lithography or by photogravure.

The processes of rubber offset printing, of machine collotype, and of photogravure have given new possibilities to map printers. Offset printing may be on a matt surface paper as in the Swedish World Atlas. Photogravure can economise by printing from one plate what would otherwise require two, as in the map of Jan Mayen Island, *Geog. Jour.* XCIV, following p. 126, where names and hill shading are from a single printing.

62. *The mounting of Ordnance Survey Maps* has lately seen many changes and some ingenious improvements devised for the convenience of the motorist rather than the walker. A particularly ingenious fold for the 10-miles-to-the-inch map is described with a diagram in *Geog. Jour.* LXXXI, 439, and briefly on p. 67 of this book.

71. From this we may derive the following table of equivalents in the two grids:

Easting		Northing	
100 km. is	671 917 yards	0 km. is	1 109 361 yards
200	781 278	100	1 218 722
300	890 639	200	1 328 083
400	1 000 000	300	1 437 444
500	1 109 361	400	1 546 805
600	1 218 722	500	1 656 166
700	1 328 083	600	1 765 527
		700	1 874 887
		800	1 984 248
		900	2 093 609
		1000	2 202 970
		1100	2 312 331

For convenient ways of numbering the grid see *Geog. Jour.* LXXXVII, 325.

72. *Maps of the Geographical Section, General Staff.* The G.S.G.S. sheets described on pp. 73–4 and many of the three series mentioned below are not now available.

Europe 1/1,000,000 and Asia 1,000,000.

Sheet lines and style of the International Map, but with some important names in the English convention. Revised about 1936 from the provisional series compiled by the R.G.S. during the war of 1914–18. See the small diagram on p. 289.

Asia 1/4,000,000.

The earlier sheets have been revised, and the map extended over Europe in two sheets—Central Europe and Mediterranean, with a new combined sheet of the Near East.

Africa 1/2,000,000.

There are eleven French sheets of N.W. Africa and two of Madagascar; only some of the former are extended to 10° of latitude. The rest of Africa is now covered by the British series.

75. *Survey of India.* Burmese sheets.

Burma. One inch to the mile: 1/63,360.

Sheet 15 by 15 minutes in the style of the Indian sheets of the same scale, but with elaborate symbols for river banks, beds, etc. Contours at 50 feet and brown hill-shading from the south-west.

Burma. One inch to four miles: 1/253,440.

Sheet 1° by 1° in the style of the Indian Degree sheets. Relief by form lines and hill-shading from the east.

77. *Dominion of Canada.* National Topographic Series. Canada. One inch to the mile. Department of National Defence, G.S.G.S.

Surveyed with air photography by the R.C.A.F. Brown contours at 25 feet; roads filled yellow, railways black, power lines red, all classified. Grid in purple, 1000 yards, on "modified British Grid system".

Other sheets on the same scale by the Department of Mines and Resources have a four-mile grid divided into quarters by dotted lines.

Canada. Two miles to the inch. Department of Mines and Resources.

Sheets about 24 by 17 inches, covering 1° by 30 minutes. Provisional sheets uncontoured. Roads filled yellow, water blue. Confused by township boundaries and concession and lot numbers. Margin divided for four-mile grid.

Canada. Four miles to the inch. Department of National Defence, G.S.G.S.

Sheets 2° by 1°. Similar in style to the above, but with fewer symbols. Contours at 200 feet. Some sheets layer-coloured in brown at 400 feet interval. Margin divided for 10,000 foot grid. Diagram of compass variation, changing 5° on one sheet.

Canada. Eight miles to the inch. Department of Mines and Resources.

Sheets 4° by 2°. Contours at 1000 feet and layer-coloured with two green tints and brown above. Spot heights in heavy black figures. A handsome map, except for poor lettering.

Margin divided for four-mile grid.

78. *Commonwealth of Australia.* Australian Section, Imperial General Staff. Australia. 1/63,360.

Australian Survey Corps, with air-photography by the R.A.A.F.

Sheets 30 by 15 minutes; Transverse Mercator projection.

Roads in black, filled orange, railways in black, both classified. Water bright blue. Relief by contours at 50 feet in pale brown, every fourth strengthened; figured in colour of contour, difficult to read; rare spot-heights on summits only.

Grid 1000 yards, reading easting and northing, each fifth line figured on face of map.

Australia. 1/126,720.

Sheets 60 by 30 minutes; Polyconic projection.

Symbols as on one-inch map. Relief by contours in orange at 50 feet interval, too close for the scale. Place names very small. Spot heights rare.

Grid lines at 1000 yards, each tenth figured in margin.

Australia has produced nine excellent sheets of the International 1/M.

78. *Union of South Africa.*

Topographical Map. 1/500,000. Director of Irrigation, Pretoria.

Sheets 4° by 4° with some extensions; Bonne's projection.

Roads red, railways black, both classified; rivers dark blue. Relief by contours at 500 feet, figured in thousands and decimal, with six tints of green to 3000 feet and brown tints above. Air information in red. Separate characteristic sheet. Series complete.

79. *Dominion of New Zealand.*

New Zealand. Four-mile sheet. Department of Lands and Surveys.

Skeleton map. Roads, railways, and rivers in black; lakes blue. No contours, some spot heights; trig. points conspicuously numbered and lettered. County boundaries in red with ribbon. Sheet 2° by 1°.

80. *Crown Colony of Cyprus.*

Cyprus. One inch to the mile, 1/63,360. Visitors' Maps.

Roads in black, classified; rivers blue. Relief by contours at 100 feet in brown, every fifth strengthened, and spot heights. Symbols in black for conifers, fruit trees, and vineyards. The regular sheets only in black photographic prints on two-inch scale.

There is also a quarter-inch sheet of the whole island, with relief by conventionalised hachures.

80. *Mandated Territory of Palestine.*

Palestine. 1/100,000. New series covering 50 kilometres square.

Roads classified, railways black, rivers blue. Relief by contours at 25 metres in red, every fourth strengthened, and spot heights, all well figured in old-style numerals and very legible. Elaborate coloured ground symbols, for vegetation, orchards, crops, and sand; rock drawing in black.

Grid 10 km. A good clear map compiled, drawn, and printed in Palestine under the direction of the Commissioner for Lands and Surveys.

85. *The state of the International* 1/1,000,000 *Map* at the outbreak of war was summarised in *Geog. Jour.* XCIV, 404, with an index showing the maps of Europe and adjacent countries and distinguishing between those that followed the International Convention strictly and those produced in somewhat varying styles. The Carte Internationale has not proved a good example of sound international cooperation. Few states have compiled the contributions of their own territory to the content of a joint sheet, and this has led to rival editions of the same sheet. There are several examples of an international map turned to national ends and there is great discrepancy in the choice of contours and of the places where layer-colouring changes. The text diagrams taken from this article show how incomplete even in 1939 was the International Map of Europe and how important is the series of Europe and Asia 1/1,000,000 published by the Geographical Section General Staff.

87. *Flying maps* are being standardised on a scale of 1/500,000. The European series has relatively little detail except special flying information, and a clean-looking system of layer-colouring in violet. They have no grid; the range of modern flight makes this impossible. References to flying maps are presumably in latitude and longitude, and the figuring of the meridians and parallels needs improving.

88. *Maps of Europe.* With the greater part of Europe enemy territory or in enemy occupation it is useless in 1941 to make any serious attempt to revise this chapter, since few of the maps are obtainable. But it may be useful to describe briefly certain maps which were omitted from the third edition; to mention some others there included which have made no progress and may be considered abandoned; and to correct one or two mistakes.

89. *Longitudes from Ferro* are now at last nearly extinct on the map of Central Europe.

91. *Belgium* 1/100,000: there is a new edition with larger sheets, about 31 by 24 inches.

92. *Czechoslovakia* published about 1936 a 1/500,000 of Central Europe in nine sheets, covering 4° by 2°, with longitude from Greenwich, brown contours at 100 m. and grey oblique hill-shading.

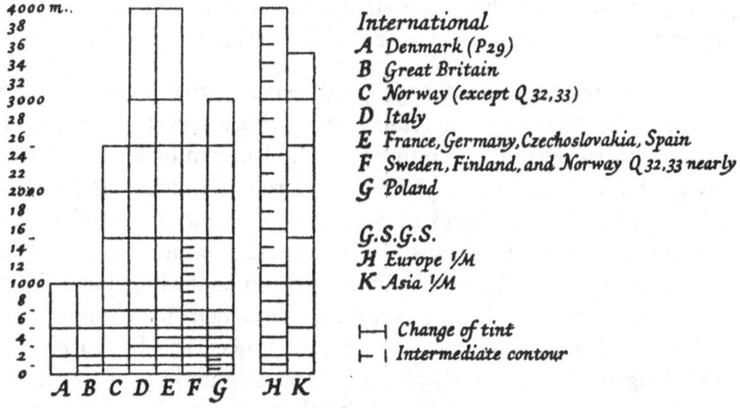

Fig. 33. *Layer colouring and intermediate contours adopted by European States contributing sheets to the International 1/M Map.*

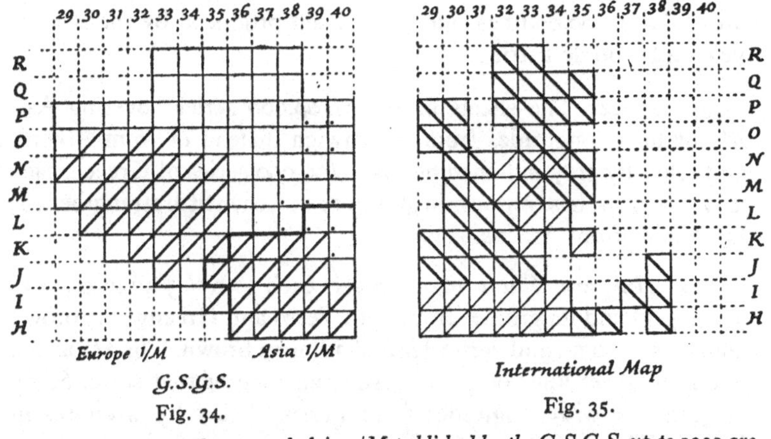

Fig. 34.

Fig. 35.

Fig. 34. *Sheets of Europe and Asia 1/M published by the G.S.G.S. up to 1939 are shown by squares with solid outline; the division between Europe and Asia sheets by a heavy line. The sheets revised and layered are distinguished by a diagonal line: those in "rough style" by a dot in the lower left-hand corner.*

Fig. 35. *Sheets of the International 1/M Map published up to 1939 are shown by squares with solid outline: those conforming to the Paris Convention by diagonal ⟍, those not conforming by diagonal ⟋.*

94. *Estonia* had produced a good 1/200,000 covering each 1½ by 1 degree, with heights in metres and contours at about 16 m. interval, presumably based on the old Russian surveys in *sajen*.

94. *Finland* had about 1932 produced a 1/50,000 series 30 by 15 minutes, with brown contours at 2 *sajen* interval, covering the south-eastern part of the country, on the Russian border. Another series at 1/100,000, 40 by 30 cm., had brown contours at 25 m., a 2 km. grid on central meridian 30° east of Greenwich, and elaborate symbols described in Finnish only. Another series on this scale, of later date, is the Economic Map, with no contours, spot heights in metres, more elaborate symbols described in Finnish and Swedish. A fourth series on 1/400,000, sheet 16½ by 19 inches, with longitudes from Helsinki, has spot heights in metres but no contours, names in Finnish and Swedish, the smaller names very small.

95. *France:* the 1/80,000 special series Frontières des Alpes has now been superseded by the 1/50,000 *type* 1922 covering the same area. The 1/100,000 has since 1922 been produced by the Service géographique de l'Armée.

99. *Greece:* The sheets of the 1/100,000 series covering Asia Minor have longitudes from Greenwich instead of from Athens, and relief by grey hill-shading. A 1/400,000 series of Greece, each 2 by 2 degrees, has contours in grey at 200 m. each fifth strengthened; longitudes from Athens.

100. *Hungary:* The 1/75,000 covers each 30 by 15 minutes, with 3 km. grid and longitudes from both Ferro and Greenwich; brown contours at 10 m. and vertical hill-shading in brown. The 1/200,000 is on the sheet lines of the old Austrian map of that scale, 60 by 60 minutes, with longitudes from Ferro. Relief by hachures in brown. The series on 1/750,000, measuring 39 by 33 cm., has elementary brown hachures.

101. *Italy:* The 1/75,000 and 1/200,000 described in the last edition made little progress and may be considered abandoned.

102. *Latvia* has published two series, on 1/75,000 and 1/200,000. The former sheets cover 30 by 15 minutes, with a 3 verst grid and

heights in *asis* which seem to be the same as the Russian *sajen* or 2·134 m. The latter series is very like the Estonian map on the same scale, and with the same contour interval of about 16 m.

102. *Norway*. The old series of *Amt* maps on 1/200,000 should have been included in the last edition. They are in black, with heights in Norwegian feet, and contours at 100 ft. interval, too close for such mountainous country. Large and awkward sheets, in skeleton outside the boundary of the *Amt*.

103. *Portugal* has also an old series 1/100,000 in black, with 20 m. contours and a 5 km. grid easily confused with the graticule.

106. *Sweden* 1/50,000: newer sheets in full colour with brown contours at 5 m.; other sheets in black with lakes blue but not rivers, and hachured. The 1/100,000 series has relief by mixed hachures and form lines, with heights in Swedish feet. The northern sheets of both have sheet lines along parallels and meridians, the southern sheets rectangular, and the numbers of the first are duplicated in the second. Though longitudes are from Stockholm the southern sheets are based on the meridian 6° west, so that the graticule on the Stockholm sheet is much inclined to the sheet lines.

108. *Yugoslavia:* the 1/75,000 series produced in Vienna for the Serbian General Staff before 1914 has not been maintained.

109. The chapter on *Foreign Maps beyond the frontiers of Europe* requires much more revision than can be given to it at the present time. In the past few years Central and South American states have made a good beginning in the heavy task of covering their great territories with maps on topographical scales. Mexico has series on 1/100,000 and 1/500,000; the Argentine Republic on 1/50,000, 1/100,000, and 1/500,000; Bolivia on 1/250,000; Chile on 1/100,000; Peru on 1/200,000; and Uruguay on 1/50,000 and 1/200,000. The French, Italians, and Spanish have produced many new sheets of Africa, and the French four series of Syria. The United States Coast and Geodetic Survey have made general maps for air-communications on 1/500,000 and 1/M; and the American Geographical Society has nearly completed its magnificent series

of Central and South America on 1/M. A few of these will be described briefly in the following notes.

109. *Egypt* has not proceeded with its 1/50,000 map, but has a new series on 1/500,000.

110. *French Africa; Algérie:* the map is on 1/200,000, not 1/100,000 as stated in the last edition.

Maroc: There are now also maps on 1/50,000, 1/100,000, and 1/200,000.

111. *Indochine:* There is a map on 1/100,000 in progress, in place of the 1/200,000, which seems to have been discontinued.

112. *Italian Africa:* The Italians had maps of Eritrea also on 1/50,000 and 1/400,000; of Cirenaica on 1/100,000; and a series including Ethiopia on 1/M.

114. Thailand (Siam) had sheets also on 1/200,000 and 1/500,000, all in Siamese.

114. *Turkey* has a new edition of the 1/200,000 with the names in the Romanised alphabet.

115. The *United States* has two important recent maps made by the U.S.C.G.S. for the Bureau of Air Commerce, Washington, D.C.

United States Sectional Aeronautical Chart, 1/500,000.

Sheets 6 by 2 degrees (half the International sheet), but on Lambert's conical orthomorphic projection with standard parallels 33° and 45°, with graticule in black carried across sheet. Names stamped in condensed sans-serif, except for towns over 100,000. Relief by contours at 1000 feet; layer colouring in two tints of green and five of brown, changing at 1000, 2, 3, 5, 7, 9, and 12,000 feet; built-up areas yellow. Railways black and highways purple, both classified; power lines and full air information in red.

United States Regional Aeronautical Chart, 1/Million.

Sheets 12 by 6 degrees. Same projection and characteristics and symbols as above, but each degree meridian and parallel divided to minutes.

118. *Atlases:* The first volume of a remarkable World Atlas and several excellent National Atlases have been produced in the last ten years, and must be described briefly.

The Great Soviet Atlas of the World, Vol. I. Moscow, 1937.

Excellent colour printing in the style of the Swedish Världsatlas by the offset process, with 83 maps of the World, physical, economic, historical, and demographic: some of unusual interest, such as the gravimetric map, general soil map, rainfall map covering the oceans as well as the land, and the map of climatic types. The remaining 36 plates show the whole U.S.S.R. in a single sheet, larger sheets of European Russia, and excellent maps of geology, glaciology, geomorphology, soils, minerals and vegetation, with 27 diagrammatic plates of economic matters. The very elaborate tables of symbols are unfortunately described only in Russian.

An Atlas of the Leningrad Region and the Karelian A.S.S.R. was published by the Leningrad State University in 1934, with titles translated into English.

Suomen Kartasto (Atlas of Finland): Geographical Society of Finland. 1925.

One of the first of the National Atlases, with four maps of Finland on 1/M and many good physical and economic maps on smaller scales.

Atlas de France: Comité National de Géographie, in course of publication from 1933 in separate sheets.

Begins with the cartography, with specimens of the different scales, and excellent layered maps on 1/1,250,000; good physical and vegetation maps, profiles of rivers, areas of floods, and the economic maps less diagrammatic than is usual: a fine atlas.

Atlas des Colonies Françaises: Société d'Editions Géographiques, Paris, 1934, interleaved with text.

Atlas of the Czechoslovak Republic: Czech Academy, Praha, 1935.

Good layer-coloured general map on 1/1,250,000, and many well designed and printed physical and economic maps with titles in Czech and French. Sumptuous and high-priced.

New Atlas of China, compiled by Dr V. K. Ting, Dr W. H. Wong, and S. Y. Tseng: Shun Pao Press, Shanghai, 1934.

Issued to commemorate the sixtieth anniversary of the newspaper Shen Pao. Political and physical maps in pairs, mostly 1/2 M, but Mongolia, Sinkiang, and Tibet on 1/5 M. Several economic and 58 city maps. Gazetteer of 180 pages containing 33,000 names, with geographical coordinates and indication where the position is determined astronomically. But entirely in Chinese.

Atlas van Tropisch Nederland. Batavia, 1938.

Produced by the Royal Netherlands Geographical Society in collaboration with the Topographical Service of Netherlands India. Thirty-one sheets on Mercator's projection and many scales. Remarkable for its maps of distribution of volcanoes and earthquake epicentres.

121. For further illustration of the graphical determination of position near the pole, see papers in *Geog. Jour.* LXXXIX, 255–9, and XC, 164, discussing the observations made by Peary at the North Pole.

127. *Wireless time signals*, especially on short wave, have multiplied in recent years and receivers have correspondingly improved, while the multiplication of high-power long-wave stations has made it increasingly difficult to hear the time signals of Rugby and Bordeaux, Nauen and Annapolis free from interference, unless one can carry the heavy Marconi R.P. 11 set with phasing unit. There are zones around short-wave stations in which the signals skip, but there are now so many stations that this matters relatively little.

144. The Bagnold *sun compass*, described in *Hints to Travellers*, I, 147, requires a programme calculated for the journey from Nautical Almanac and Azimuth Tables. The Survey of Egypt has produced the Evans-Lombe sun compass cards for belts of latitude four degrees wide and with thirteen cards in each series to allow for change of the sun's declination with date. These may be considered superseded by the Cole sun compass based upon the theory of Richards and Cole given in *Geog. Jour.* XCVII, 351–3. Here a single dial with a gnomon movable along the meridian provides a universal instrument to take the place of the latitude series with thirteen cards in each.

152. *Telescopic alidade.* With improved construction of instruments opinion has changed and the best explorers now prefer a plane-table outfit with telescopic alidade such as Wild's.

174. *The Durand reconnaissance ladder* is now some fifty years old, but perhaps now superseded by the use of easily erected towers, both for reconnaissance and for the triangulation: see the reports of United States and Canadian Surveys which make much use of them.

176. For a light and easily operated *electric beacon* see the description of the Danish Geodetic Institute pattern (*Geog. Jour.* XCI, 397).

177. In suitable country such as Canada survey parties are now generally moved and supplied by air.

179. *The triangulation of Great Britain* has recently been replanned and reobserved and all the new stations are permanently marked.

181. *Astronomical positions.* The zenith telescope is not now in much use except perhaps by the Survey of India. The prismatic astrolabe in its improved form as designed for the British Hydrographic Service (see p. 268) gives excellent results, but in common with all methods that require observation at a fixed altitude, wastes much time in broken weather. It is often preferable to preserve the quadrantal method of observing stars near the middle points of the four quadrants, but to observe with a theodolite instead of the astrolabe, which gives a much wider choice of stars. Each observation gives a position line, and the distribution of the position lines when plotted gives immediate check upon the accuracy of the results.

The best modern practice is to multiply the number of astronomical positions and to give less time to each.

183. *Wireless time signals.* The pattern of the international and the rhythmic time signals remains unchanged but the number of signals is greatly increased and the R.G.S. pamphlet is obsolete. Short-wave signals are especially multiplied, the apparatus for receiving them much improved, and they are now generally employed. There are so many that the older difficulty that they have zones of inaudibility is no longer serious. One or another may almost always be heard.

206. *Survey for artillery.* These paragraphs are now mainly of historic interest. It is not possible to describe present practice, and the conditions of defence in depth are entirely different from those of static warfare.

208. *Sound ranging in water* is now much used in hydrographic survey. For an account of recent methods of the United States Coast and Geodetic Survey, see the paper by Professor O. T. Jones on submarine canyons, *Geog. Jour.* XCVII, 80, 185.

217. The observations of Maupertuis in Lapland made the flattening of the earth much too great, and if the error had been of the same size but of opposite sign his result would have been contrary to the truth, so that he was fortunate. (See *Geog. Jour.* XCVIII, 291, from which Fig. 36 is taken.)

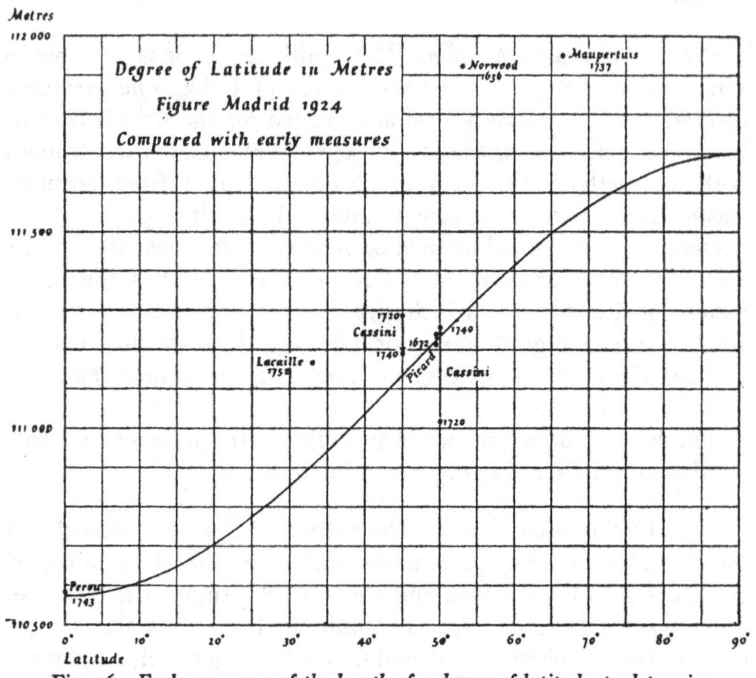

Fig. 36. *Early measures of the length of a degree of latitude, to determine the Figure of the Earth.*

221. More recent work by Dr Meinesz and by others who have followed his method has greatly extended the range of gravity survey at sea, and the accuracy of the instruments has been so much improved that systematic error of the results is now due largely to the difficulty of determining the correction for sea currents.

223. *Isostatic compensation.* It is now generally held, except in the United States, that there is no justification for assuming that isostasy is complete at any particular depth. See discussion on

"Problems of the Earth's Crust", *Geog. Jour.* LXXVIII, 438–55, and the reviews of Meinesz' work in *Geog. Jour.* LXXXII, 444, LXXXV, 479.

227. *Geodetic bases.* Professor Bonsdorff proposes that a few standard bases should be measured with all available accuracy in Europe and that the geodetic tapes should be standardised upon them instead of on short comparators in the laboratory. (See *Reports* of the Finnish Geodetic Institute and *Proceedings* of the Baltic Geodetic Commission.)

228. *Prismatic astrolabe.* Observers are no longer unanimous in preferring this instrument to the theodolite, partly because it is more important to have many astronomical stations of moderate accuracy than a few with an apparent accuracy out of proportion to the probable deviation of gravity at the station. At the Stockholm Geodetic Conference 1930 it was alleged that the astrolabe gives errors 50 per cent greater than other instruments for determining local time. See, however, review of the Geodetic Reports Survey of India 1936 and 1937, *Geog. Jour.* XCIII, 69, which contains a good account of recent geodetic methods.

238. *British triangulation.* The triangulation has been re-observed recently, but the discussion of it is not yet published. One understands that it is to be used for correcting the old triangulation network, leaving it based upon Airy's figure of the earth instead of making a completely new reduction upon a more modern figure.

244. *The photo-goniometer* requires that the optical conditions in which the plate is taken must be reproduced in the measuring machine, using either the objectives by which the plates were taken or others optically identical. This severe limitation has now been abolished in Dr Heinrich Wild's Autograph A 5 which can measure any plates within a wide range of focal length. Modern practice as exemplified in the Stereographic Institute of Professor Schermerhorn at Delft restricts the use of the large stereo-plotting machines such as the stereoplanigraph or the new Wild autograph to air triangulation. The plotting of detailed contours is done upon simpler machines such as the Zeiss Simplex when the constants of the plates have been established.

262. *Bubble sextants.* The demands of air navigation have greatly improved these instruments. The *Air Almanac*, published quarterly, and *Air Navigation Tables* have much improved the speed and accuracy of astronavigation. (For the *Air Almanac* see *Geog. Jour.* XCII, 133.)

One does not yet hear of bubble sextants being used on land, but they would seem well adapted for desert navigation.

270. *Geodetic pendulums* of the von Sterneck pattern are now obsolete in British use, superseded by the Cambridge pattern based upon the Meinesz design. For recent methods see papers on Gravity Survey in Africa, *Geog. Jour.* LXXXIX, 91, and a memoir in *Phil. Trans.* XXXV, 445–531, by Dr Bullard. For more recent static pendulums devised by Father Lejay and others, see *Geog. Jour.* LXXXV, 198, and XCIV, 92. Observation with these instruments is very rapid but they are difficult to standardize and are used principally for detailed survey about stations at which gravity is determined with the moving pendulums.

272. *Surveying cameras.* Modern designs have aimed at reducing weight both of the instrument and by substituting films for plates. Much can be done by adapting small precise cameras such as the Leica carried by a gallery upon a small mountain theodolite. A very beautiful but much too expensive miniature photo-theodolite was produced by Messrs Zeiss about 1938.

Chapter XV

FURTHER ADDITIONS TO PRECEDING CHAPTERS

13. *Boundary delimitation and demarcation.* When a new international boundary is to be made it is first delimited in a diplomatic convention; then demarcated by a boundary commission; and finally described in the report of the commission, with list of positions of the boundary pillars and a map of the boundary neighbourhood.

Delimitation is the geographical definition of the line which is to be marked. In country which is settled and mapped it will be defined as following natural features, watersheds or rivers; or existing internal boundaries, of provinces or communes; or as a line to be fixed leaving certain villages on one side and certain on the other. In country which is unmapped and sparsely inhabited it is too often defined geometrically, as a meridian or parallel or a straight line joining two points, without consideration of the difficulty of determining and marking such lines. In either case there should be a reconnaissance of the frontier zone, if only by air, before the delimitation convention is drafted, to avoid major mistakes. One must remember that astronomical determinations of latitude and longitude are affected by local deviations of the vertical; that geodetic latitudes and longitudes are affected by the mean values of such deviations over the area triangulated, and by the values adopted for the figure of the Earth, that is to say, for the semi-axis major and the flattening; and that there will be systematic differences between the coördinates derived by two contiguous but independent surveys. One must avoid definitions which may be self-contradictory, such as the source of a river or the junction of two rivers in lat. —, long. —; must avoid saying in one place the coast and in another the sinuosities of the coast, which might imply that the former meant the generalised coast line. One should indeed be careful of using the word coast at all, which in Alaska means the seaward edge of the land, and in Labrador all the country to the "height of land".

The treaty of delimitation should give the demarcators power to agree upon small deviations from the letter of the treaty: not compel them, for instance, to erect a boundary pillar inside a house because pillars are to be half a mile apart; should give them power to omit a pillar which comes in an impossible place such as a precipitous ice-slope on the Alaska boundary, but require them to mark minutely where it crosses streams on the same boundary which may be full of gold. The common provision that if the text and the map are discordant the text shall prevail becomes ineffective when the map is so bad that the text based upon it is meaningless, as happened on the Lake of the Woods section of the United States-Canada boundary.

Some of the more celebrated cases are discussed with sketch maps in *Geogr. Jour.* Dec. 1921, LVIII, 417–43. We may look briefly at the principal difficulties.

Uganda-Congo-Tanganyika (German East Africa).

A declaration of 1885 made much of eastern boundary of Congo the meridian 30° E. Convention of 1890 made neighbouring boundary between British and German E.A. the parallel 1° 20′ S., with diversion south if necessary to include "Mt Mufumbiro" of Speke in B.E.A. Ruwenzori, lakes Edward and Kivu, and the Virunga volcanoes then unknown. Geometrical boundaries proved entirely unsuitable when this magnificent country of the Western Rift came to be explored, and after long surveys and negotiations they were greatly changed at Brussels Conference of 1910. A striking case of unsuitable delimitation.

Alaska Boundary.

Delimited by Anglo-Russian treaty 1825 and in nearly same words when Russia sold Alaska to United States in 1867. Boundary to be up Portland Channel to 56° N., then crest of mountains parallel to the coast to meridian 141° E. If this crest more than 10 marine leagues from the ocean, a line parallel to the sinuosities of the coast.

The channel had three navigable mouths between islands, and forked south of 56° N.; whole coast studied with islands, with deep bays and fjords. Is coast the limit of open sea, or of all territorial waters including deep inlets? Or is it the "envelope" of the shore-line? Does ocean include territorial waters inside islands? When whole mainland mountainous, what is the crest, and what is a line parallel to the sinuosities of the coast? These and minor questions decided by arbitration 1903. A clear case of inexact delimitation.

Argentine-Chile Boundary.

Delimited by Treaty 1881 and Protocol 1893. Protocol said: All land and all waters east of the line of most elevated crests of the Cordillera de los Andes that divide the waters shall be Argentine....Sovereignty over all territory east of the principal chain of the Andes to the Atlantic coast.

But the highest crests often outliers of the principal chain, and the watershed often well out on the Argentine pampas; moreover, there are basins of internal drainage.

Dispute referred to arbitration of Great Britain. Tribunal sat in London 1899 and directed survey of the ground by British Commission, which reported to tribunal, and Award made in 1902, to best interpret the intention of the contradictory delimitation.

South Australia-Victoria.

An Act of 1834 empowered William IV to erect South Australia into a British Province between the meridians 132° and 141° E. A point about 141° E. on the south coast was determined in 1839, by transport of chronometers from Sydney. From this in 1847 southern end of meridian marked and extended north, eventually to the Murray. This terminal later found 2 miles 19 chains west of the meridian. South Australia claimed rectification. Case came on appeal to Judicial Committee of the Privy Council in 1913. Judgment in 1914 that a boundary surveyed, marked, and proclaimed must stand.

Cases for revision of faulty demarcation are uncommon. Here most of the error due to error in assumed longitude of Sydney.

Canada-United States (Lake Superior to Pacific).

Treaty of Paris 1783, based on Mitchell's map of 1755, made the boundary run through Lake Superior to the Long Lake, across its middle, and by the water communication to the Lake of the Woods, to the most north-western point thereof, and thence on a due west course to the Mississippi.

It was soon found that there is no water communication; that in tangle of islands and inlets north-west corner indeterminate; and that Mississippi rises south of a parallel through any point of the lake. Boundary on parallel 49° N. to Rocky Mountains agreed by Convention 1818 and extended to Pacific by Treaty 1846.

Parallel 49° N. to Rocky Mts determined by astronomical latitudes about every 20 miles. Deviation of vertical large; greatest deviation between adjacent stations 7″.98 or 750 feet; hence line marked not a parallel, but line between each pair of pillars given the curvature of the parallel.

Parallel 49° west of Rocky Mts marked 1858–61, by astro. latitudes about 14 miles apart. Again large deviations of vertical, once eight seconds in nine miles. Here a mean parallel marked over greater part, after parallels cut east and west from adjacent stations had passed one another hundreds of feet apart; the rest by great-circles between pairs of pillars.

A good example of necessity for geodetic triangulation in advance of geometrical boundary making.

Maps

65. Great Britain. One inch to the mile, 1/63,360. Second War Revision 1940. Geographical Section, General Staff, No. 3907. Published at the War Office 1941. Put on sale to public 1943.

Engraved surface of most sheets 27 by 18 inches; sheet-lines of the Ordnance Survey one-inch Fourth (Popular) edition. From about Birmingham northwards the style is that of this Fourth edition, which is on a Cassini projection; southward it is that of the Fifth edition, which is on a Transverse Mercator projection. Each margin bears the words W.O. Cassini Grid, which are hard to interpret, since a grid belongs to the rectangular coordinate system on which projection and map are drawn, and not at all to the projection itself. The grid-like lines on this map, at the equivalent of one kilometre interval, drawn at a slowly varying obliquity to the sheet-lines, must be what is called on page 71 "something very like a true grid". Nothing is said of central meridian or origin; but it is not the new National Metric Grid of the map on 1/625,500 published in 1942, which will be shown on Ordnance Survey one-inch sheets after the war. The W.O. Cassini Grid may therefore be considered as not intended for civil use.

67. Great Britain. 1/625,000, about 10 miles to one inch.

In two sheets 40 by 32 inches. Ordnance Survey 1942. Outline, railways, and main roads in black.

Transverse Mercator projection on National Meridian 2° west. Origin 49° north; false origin 400 km. west and 100 km. north of true origin. Ten kilometre grid: the first representation of the new National Metric Grid to be put on sale. (See pages 70, 71, 285.) This map to be used as base for distribution maps prepared by the Ministry of Town and Country Planning.

103. Polish Military Geographical Institute. Mapa Polski i Krajów Ościennych (Map of Poland and neighbouring Lands). 1/1,000,000.

Four sheets 27½ by 22½ inches, overlapping so that Warsaw is in a corner of each. Conical projection on central meridian 21° E.

Contours at 100 metres (dotted), 200, 400, 600, 800, 1000 (strengthened), 1400, 1800, 2200, 2600. Layers in shades of green, yellow, brown, and violet. All names in Polish, e.g. SZCZECIN for Stettin; all black names in sans serif.

Printed by Bartholomew, Edinburgh, 1943.

78. Map of Antarctica, with Handbook and Index. Canberra, Commonwealth of Australia, Department of External Affairs, 1939.

In two sizes: a single sheet with engraved surface 36½ by 27¾ inches, polar azimuthal equidistant projection, latitudinal scale 1/10 M, with large inset of the Australian territory on 1/7M; and in two sheets with a small overlap on 1/7½M from the same drawings. Printed in eight colours, the inland topography, mountains and glaciers well drawn. Heights and soundings in metres. Boundaries of political sectors and subsidiary divisions into Lands and Coasts shown in red. Bathymetric tints of the sea nicely graded.

Excellent Handbook by E. P. Bayliss, who compiled the map, and J. S. Cumpston, with historical introduction and summary of work since 1925. (Fully reviewed in *Geogr. Jour.* Dec. 1940, XCVI, 435–8.)

115. Antarctica. Chart No. 2562, Hydrographic Office, Washington, D.C., March 1943.

Sheet 47 by 30 inches, polar equidistant projection, latitudinal scale 1/11,250,000, printed in six colours. Graticule in brown on separate plate from outline. Buff tint on land seen more or less by at least one person from the air. Sea contours at each 500 fathoms, layer coloured beyond the 100 fathom line in ten tints of blue. Land relief by elementary drawing of mountain ranges, with many heights of peaks, but no spot heights on plateau or at Pole. No indication of territorial claims to Antarctic lands, as none recognised by the Department of State, but names associated with individual claims by U.S. citizens given prominence. Sailing Directions for Antarctica, H.O. 138, since published (Chart fully reviewed in *Geogr. Jour.* July 1943, CI, 29–33. Sailing Directions Jan. 1944, CIII, 77–9.)

282, 283. Europe and the Middle East. 1/11,000,000. Made for the British Council by the Royal Geographical Society.

Sheet 38 by 25 inches, printed in photogravure in six colours. Murdoch's Third Conical Projection, proposed in 1758, and first calculated, for this map, in 1940 (*Geogr. Jour.* June 1941). Contours at 50, 100, 200, 500, 1000, and so by thousands of metres. Layer-coloured with twelve tints of buff on one plate, possible only by photogravure. Delicate oblique hill-shading in lavender grey. No border, but margin of one inch to give room for figuring graticule, scale values of parallels, and occasional information.

Drawn for reproduction same size, quill-pen lettering throughout, first time on so important a map. Names relatively few, to avoid masking representation of relief, but about 2100, with 1100 contour and spot heights.

Principal railways by single red line; important roads, especially where no railways, and main desert routes all in dotted red line; town signs also in red. This delicate web of red over highly developed countries distinguishes them from the undeveloped, but scale too small to make a real guide to roads and railways. Political boundaries, as before 1938, in black dotted line, and names of States as before enemy aggression in that year.

Printed on very strong paper, half linen, half cotton, which passed stringent tests for folding and tearing; mounting on fabric not really necessary. Flat sheet 38 by 25 inches folds into twelve in two rows of six, 12½ by 6½ inches.

A full account of the novel features in design and production of this map in *Geogr. Jour.* September 1942, C, 123–130. See Plate XXVII.

291. Hispanic America. 1/1,000,000, in style of the International Map. American Geographical Society, New York.

The first sheet, La Paz, published 1923, and all but the last three of the 104 sheets received by end of 1943. The sheets cover all America south of the northern border of Mexico, and include the West Indies: a remarkable contribution to the 1/M map of the world.

Map of the Americas. 1/5,000,000. American Geographical Society, 1942.

Three large sheets, of engraved surface 43·4 by 29·7 inches, cover the whole area of Hispanic America represented by the 104 sheets on 1/M now almost complete, and extend into the United States up to New York on the Atlantic Coast. Contours in brown at 200, 500, 1000 metres and so by thousands on land and in the ocean bed, with an improved scale of layer colouring on land changing at each contour, and another colour scale for depth of sea changing at every second contour below 200 metres. Contouring in sea complete and very detailed up to Nantucket, but no relief of the land north of the Mexican border.

Railways black. Pan-American Highway, all-weather, dry-weather, or projected roads, in four styles of red symbol; other improved roads and proposed Federal highways in two more symbols. Rivers, glaciers, marsh, and canals in blue; salinas blue and black symbol. Airports, principal landing grounds, and hydroplane anchorages in red. International and major civil boundaries classified.

Populations of towns shown by six styles of writing names, and civil importance by three symbols. Drawn on a new oblique bi-polar conical orthomorphic projection, with full inset notes on measurement of distances and areas, table of scale conversions, etc. Glossary of abbreviations and translations of Spanish and Portuguese topographical terms. Full referenced index of names in separate pamphlet.

To be extended to cover the whole of North America.

World Maps made in the Carthographic Department of the National Geographic Society, Washington.

The World in two hemispheres. Scale at centre 527 miles to the inch. Zenithal equal-area projection. Insets land and water hemispheres, time zones. Layer coloured in buff at 200, 500, 1500, and 3000 metres. Dec. 1935.

South America. 1/8,500,000. Equal area polyconic projection. Insets, airways and relief 1/28M. Dec. 1937.

Central Europe and Mediterranean. 1/5,000,000. Conical projection with two standard parallels. Oct. 1939.

Mexico, Central America, and West Indies. 1/5,702,400. Conical orthomorphic projection. Enlarged insets. Dec. 1939.

United States. 1/5,195,520. Albers' Conical equal-area Projection. Insets Populations of States (Census of 1940), and National Defense. Dec. 1940.

Indian Ocean. 1/20,000,000. Mercator's projection. March 1941.

Atlantic Ocean. 1/20,000,000. Stereographic projection. Historical notes in sea. Sept. 1941.

North America. 1/12,000,000. Zenithal equal-area projection. May 1942.

Theater of War in Europe, Africa, and Western Asia. Zenithal equidistant projection. Layer coloured. July 1942.

Asia and adjacent areas. 1/17,500,000. Transverse polyconic projection. Table of air-line distances. Dec. 1942.

Africa. 1/11,721,600. Zenithal equal-area projection. Inset relief, table air distances. Feb. 1943.

Northern and Southern Hemispheres. 1/40,000,000. Zenithal equidistant projection. Some main results of U.S. Antarctic Service expedition. April 1943.

Europe and the Near East. 1/6,000,000. Zenithal equidistant projection. Inset Middle East. Air-line distances. Table of English and National names of countries. June 1943.

Pacific Ocean. 1/27,500,000. Mercator's projection, Enlarged insets of island groups. Sept. 1943.

The World Map. 1/40,000,000. Van der Grinten's projection. Polar regions in spandrels. Dec. 1943.

Most of these maps have a ground tint of yellow, and political boundaries with tinted ribbon. Relief by simple hachures. Some later sheets have names from labels printed in new well-designed characters. The maps are published with the *National Geographic Magazine*; may be bought flat or mounted on linen only by members of the N.G.S.

116. *Great-Circle Courses and Distances.* Civil aviation and radio engineering demand world-maps on which one may lay out great-circle courses and measure great-circle distances. Charts on the gnomonic projection have been made, especially by the U.S. Hydrographic Office. The straight line joining any two points on the gnomonic chart is a great-circle; but gnomonic projections cannot be made to cover so much as a hemisphere, and it is tedious to measure distances upon them. Gnomonic charts have therefore only a limited use.

Alternatively, the graticule of meridians and parallels for a zenithal equidistant projection centered upon any chosen station may with considerable labour be calculated and drawn to cover the whole sphere. A straight line drawn from the centre to any point gives the great-circle course from the central station to that point, and the distances along those radial lines are true. But on such a projection the point antipodal to the centre becomes the whole limiting circumference, and all its neighbourhood is so greatly distorted that it is impossible to draw the outlines of that part of the world except at great expense of time, and very roughly. If the antipodal point is in the sea the effect is not so bad; if on land, that land is expanded into a ring which includes all the oceans and the rest of the world (Fig. 37). Making such a map is wasteful, because it serves only the central station.

Great-circles which do not pass within 20 degrees of the poles may be plotted very easily upon the Normal Mercator Projection (N.M.) with the aid of a Transverse Mercator (T.M.) plotted to the same scale. On a piece of tracing paper trace the equator of the normal projection and the points *A*, *B* to be joined by a great-circle. Lay this tracing on the T.M. with the trace of N.M.'s equator lying along the central meridian of T.M., and slide it so that the points *A*, *B* lie on a meridian of T.M. or proportionately between a pair of meridians. With French or Yacht Curves, or by sketching, trace the interpolated meridian through *A*, *B*. This is a great-circle on T.M. and will also be a great-circle when the tracing is replaced on N.M. Trace it through on to N.M. with carbon paper.

If the T.M. graticule is calculated with parallels at 5 or 10° interval, the length of the great-circle may be read off in degrees. But, since the distance from pole to equator is very nearly 10,000 km., it is better to calculate the parallels at 1000 km. interval, number them straight through from 0 to 40, and estimate kilometre readings at *A* and *B*. The difference between the readings will be the distance *AB*.

On the Normal Mercator the scale value at latitude 60° is twice and at 70° three times the scale at the equator, distortion is very great, and outlines difficult to draw. It is useful to calculate some Oblique Mercators, in which a great-circle passing say 35° from the poles is taken, instead of the equator, as central axis. Renumbering the meridians one can plot a series of obliques giving true but unfamiliar aspects of the world, on which any desired route lies

1. *Contours in brown.*

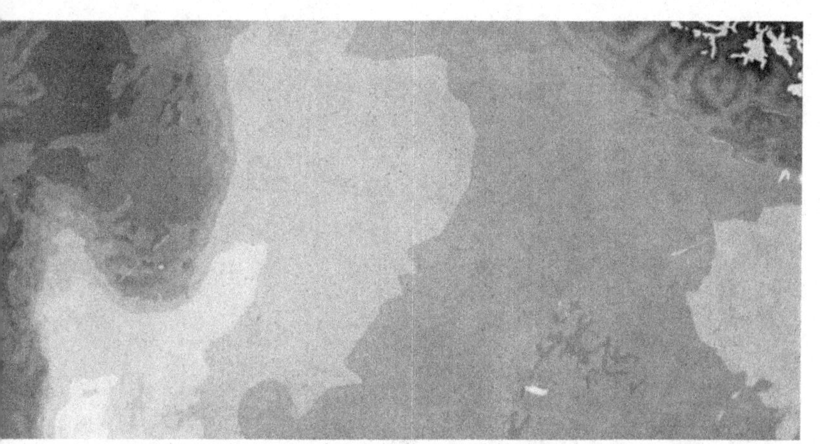

2. *Layer colours in buff.*

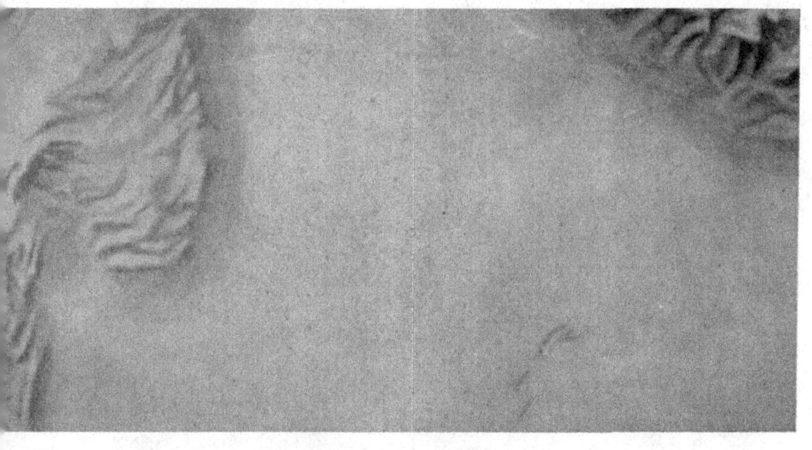

3. *Hill-shading in lavender grey.*

Parts of Drawings for colour plates, British Council Map.

Plate XXVIII

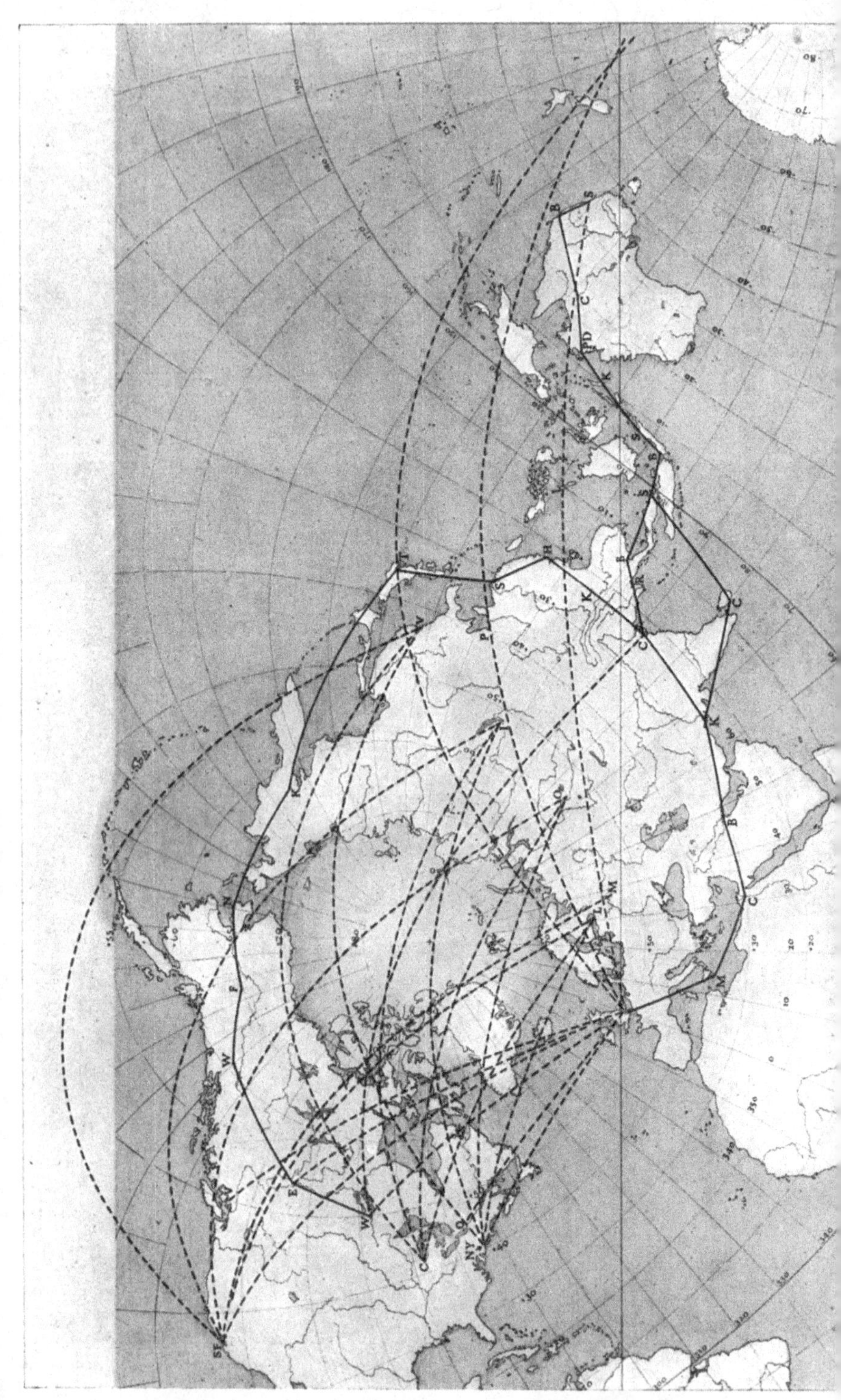

reasonably near one or other of the central axes. Great-circles and distances are found by using the T.M. graticule as before. Pl. XXVIII.

Fig. 37. *Sketch of a Zenithal Equidistant Projection of the whole World, Centre Valparaiso.*

With such a set of maps: Normal Mercator, several Obliques, and the Transverse graticule one may study all great-circle problems: compare the whole single, direct great-circle course with the more feasible and economic compound course between stations providing rest, fuel, and intermediate traffic. Each leg of this compound course may be quickly plotted and measured as a shorter great-circle the distortion along the course is minimised, and at the antipodal point is exactly the same as at the initial station.

So far one has drawn great-circles from one point to another. It is easy, instead, to lay out a system of great-circles radiating at equal differences of azimuth from any chosen station: the equivalent of the zenithal equidistant. If a great-circle is drawn from a point A in lat. ϕ_1 on true bearing θ, it cuts the two meridians 90° from A in lat. $\pm\phi_2$, given by $\tan \phi_2 = \cot \theta \sec \phi_1$. Calculating ϕ_2 for each 10° of θ we have a series of points B on the two meridians to be joined with A and its antipodal point by great-circles in the usual way. As each of these great-circles is determined mark off each 1000 km. from A and sketch the 1000 km. curves. One has then the figures to plot a zenithal equidistant projection with centre A, and the content of each compartment 10° by 1000 km. on the oblique Mercator will be the same as the content of each corresponding compartment on the zenithal equidistant. The outlines contained in one may thus be sketched into the other; Fig. 37 was drawn in this way. But the representation on the oblique Mercator is, one may think, always better than the greatly distorted version of the world on the zenithal equidistant. For reproductions of Oblique Mercators and Transverse Graticule see *Geogr. Jour.* XCVII (1941), 353–8 and folding map.

298. *Bubble Sextants on land.* Recent experience has shown that the airman's type of bubble sextant does not stand up so well as a theodolite to the vibration of a car in desert journeys. But those who have motor transport would in any case prefer a theodolite, and the bubble sextant may still be preferred by those who must travel light.

INDEX

For Additions to Index (*Fifth Edition*) see next page.

ADDITIONS TO INDEX (*Fifth Edition*)

Printed in the United States
By Bookmasters